Health and Wellness

Available *NOW* without a Prescription!

Health and Wellness
Available *NOW* without a Prescription!

Dr. Greg S. Tomalin, D.C.

Dr. Greg S. Tomalin Publishing
13606 Xavier lane Unit D
Broomfield, CO 80023
720-887-0624
303-919-5477

Table of Contents

Jaden, I see you as you really are, powerful, sensitive, determined and gracious. I see you achieving everything you choose to achieve. You can accomplish ANYTHING. I love you Buddy!

"My father gave me the greatest gift anyone could give another person, he believed in me." - Jim Valvano

Kaiya, I don't need anyone to tell me about heaven, I look at you and I believe. I love you.

"Certain is it that there is no kind of affection so purely angelic as of a father to a daughter. In love to our wives there is desire; to our sons, ambition; but to our daughters there is something which there are no words to express." ~Joseph Addison.

Acknowledgments

God has a way of teaching you life lessons.

I would like to thank my son, Jaden, and daughter, Kaiya, for showing me what unconditional love is and teaching me the meaning of life.

It's all about family. I want to thank my brother, Kevin, and my niece and nephews, Anik, Ryan and Collin, and my sister, Kari, who continually love and support me. I can truly say this couldn't have been possible without you. To the Kelley family, thank you for showing me what it means to be a family—it gives me hope for the kids.

You know you have lifelong friends when ... Making a long-distance phone call $2.00. Sending a birthday card $6.00. Flying home for a weekend $400.00. Not doing any of it and still having friends ... Priceless.

To my Canadian brothers, you are more than friends, you're family—*Sons of the Knights of the Wala Wala* for life.

Thank you to everyone on my teams at Health and Wellness Chiropractic Center's over the years (Broomfield/Lakewood/Westminster, Longmont CO) for everything you have done for me and for teaching me the power of teamwork. To all the doctors who have supported HWCC past and present, especially Drs Luneburg, Quashnick, and Gelinas, thank you for your help in editing and contributing parts to this book. Thank you, Lynda, for everything you do for me. I don't know what I would do without you. You are so good to me and the HWCC patients. Thank you for demonstrating the exercises even though you were reluctant to do so. You are truly amazing and bring balance and peace to my life.

Wendy, there is a place in my heart where your love will remain, your whispers echo and our dreams still linger. It's the place where a part of you will forever be a part of me. It has been said that you can't ever promise someone you won't ever hurt them because at one time or another you will. The real promise is if the time you spend together will be worth the pain in the end.

I have heard Anthony Robbins say, "You will only be as successful as the expectations of your peer group." I want to thank all my friends for their never-ending support. And to everyone on the Missing Lynx hockey team, thank you for keeping balance in my life and making sure I have fun at least once a week.

And finally, to all my patients, thank you for your friendship, support, and spreading the message of chiropractic. On behalf of the team at HWCC, we love and appreciate all of you.

Yours in Health,
Dr. Greg Tomalin, D.C.

Introduction

A Doctor's Confession

There isn't a day that goes by that I don't think of my mom. On the good days, I remember her before she was diagnosed with cancer. The few short years that followed her diagnosis were the toughest of my young life: the hospital trips, the drugs and surgeries, the chemotherapy—and the funeral. A year later my dad had a heart attack and was tested for arteriosclerosis in the same year my brother had part of his colon removed due to Crohn's disease. I didn't understand why this was happening to my family. Was disease and sickness just random chance or was there something more? At that time I made a decision to find out all I could about health and wellness.

The World Health Organization defines health as "a state of complete physical, mental, and social well-being, not merely the absence of disease or infirmity." Health is not merely the absence of disease any more than wealth is an absence of poverty. Health is not simply feeling fine, because problems may progress for years without causing any symptoms, as heart disease and cancer did with my mom and dad. Health and wellness can be considered holistic homeostasis of mind, body, and spirit. Health is about providing your body with something it needs for proper cell function (but currently lacks) and avoiding anything that causes toxicity. We need to live as naturally as possible.

> Drugs never cure disease. They merely hush the voice of nature's protest and pull down the danger signals she erects along the pathway of transgression. Any poison taken into the system has to be reckoned with later on even though it palliates present symptoms. Pain may disappear, but the patient is left in a worse condition, though unconscious of it at the time. —Daniel Kress, M.D.

It's a strange paradox that in a world that focuses so much on fighting disease and prolonging life, so little attention is paid to the simple, inexpensive, and powerful ways to live a life of health and wellness.

The key to health is to become proactive and not wait until you hurt before you do something about your health. The answer is not to spend more money on expensive medical tests or procedures, or to consume more prescription drugs, but rather to change how you think about health and disease.

In the vast array of choices in health care today, it is sometimes hard to know what to believe or what to think. New and expensive medical tests and procedures are introduced each year that do very little to improve health. Revolutionary diet and exercise programs emerge on an almost daily basis promising health and happiness.

Carefully crafted pharmaceutical ads show how happy you could be if you simply took this drug or that drug. Do you realize that the United States has only 5 percent of the world's population, yet we consume almost 75 percent of the world's supply of prescription drugs? You would think that if drugs were the answer to health, we would have the healthiest nation in the world, but that is not the case. The *Journal of the American Medical Association* (JAMA) and the World Health Organization (WHO) ranked the United States 37 out of 39 industrialized countries in health care (*Newsday* June 2000). In the same year an article in *JAMA* by Dr. Barbra Starfield, M.D., M.P.H., of Johns Hopkins showed that medical errors may be the third leading cause of death in the United States.

The report shows there are 2,000 deaths/year from unnecessary surgery; 7000 deaths/year from medication errors in hospitals; 20,000 deaths/year from other errors in hospitals; 80,000 deaths/year from infections in hospitals; 106,000 deaths/year from non-error adverse effects of medications—these total 225,000 deaths per year in the U.S. from iatrogenic causes, which ranks these deaths as the No. 3 killer behind heart disease and cancer. *Iatrogenic* is the term used when a patient dies as a direct result of treatments by a physician, whether it is from misdiagnosis or adverse drug reactions. This is equivalent to a death every four minutes. In fact, the general population is getting heavier, and the rates of diabetes, heart disease, and cancer are rapidly rising, and we have one of the highest infant mortality rates in the civilized world, behind the entire European Union and Cuba, South Korea, Singapore, Aruba, Greece, and the Czech Republic.

Today more than ever, tens of millions of Americans are seeking alternative opinions on how to treat and prevent disease. What accounts for this major paradigm shift? Medicine has not lived up to its billing as the be-all and end-all in health care. Traditional medicine does some great things, don't get me wrong, especially when it comes to treating emergencies, life-threatening trauma, and surgical repair of damaged body parts. My son, Jaden, is alive today because of an emergency C-section. But when it comes to managing chronic degenerative conditions, traditional medicine falls short.

One reason for this is that traditional medicine does not focus its attention on the prevention of disease. Rather, its focus is on trying to treat disease once it occurs. Many times this treats only the symptoms and never really addresses the cause of the problem. Most people in Western civilization have become high-stressed and fast-paced. Our bodies are overloaded with toxins and deficient in nutrients. We have lost step with nature. We eat unnatural foods and don't get enough exercise, and when our body begins to break down, we attempt to manage it through prescription drugs.

Wellness chiropractors take a very different approach from that taken by medical doctors. We try to educate people to prevent disease instead of treating it after it appears. If you work toward a state of optimal health, you will feel better, and you will have more energy and a higher quality of life. You will also avoid most of the expense and pain associated with the downward spiral of disease and dependence on prescription medications.

Wellness chiropractors are more than just ache and pain doctors. Chiropractic care is based on the scientific fact that within every living thing there is an inborn, innate wisdom always striving for optimal health. The body has a tremendous capacity to heal itself if it is allowed to do so. My job is to identify and remove the toxicities and deficiencies that prevent the body from being healthy. To accomplish this, I employ a variety of techniques, such as chiropractic adjustments, stretches, exercises, traction, and lifestyle changes.

My purpose for writing this book is threefold. First, because there is so much conflicting information out there, I want to present the most important concepts that are important to your health in an easy-to-understand way, backed by research. Most people will make healthier choices and live in a healthier way if they know what to do. Second, my hope is that this book can serve to help those of you who suffer needlessly because you don't understand what chiropractic is and how it can help you, including many medical doctors who still buy into the common myths surrounding chiropractic care. Third and most important, I am writing this book to help you live a healthy life, free from pain and disease, so you and your children don't have to go through what I went through. If my family had had this information and acted on it, their destiny could have been different.

Each chapter covers important topics that are relevant to your health. All chapters will be referenced with peer-reviewed scientific journals to prove without question the facts of health and wellness. There is a lot of information floating around that is just false. How many times have you heard *They say you should...* or *I heard it's good if you* ...I was taught by a mentor of mine to ask "Where did you read that? Where can I get a copy?"

In the first chapter, you will learn the basics of body mechanics and how the body works. This may be the most important chapter of all, because the better you understand how your body works, the better health choices you can make. The second chapter introduces you to chiropractic care and how it can be used to improve body functioning and improve your health. The remaining chapters contain information about nutrition, exercise, stress reduction, eliminating toxins. Throughout the book, you will find exercises, stretches, and other things you can do at home to help improve homeostatic cell function, speed your recovery from illness and injury, and keep your body healthy.

The three main ideas that I want to drive home in this book are that

- health is not merely the absence of disease

- the body is a series of complex systems that rely on proper balance, coordination, and motion in order to function correctly

- the chiropractic lifestyle is the most effective way to remove the obstacles that prevent your body from expressing its full health potential.

We all know we should change our habits to improve our lifestyles. The reason we start and fail is that we haven't made the decision to make the change a *must* and not a *should*. I hope this book shows you that you can and must acquire healthy habits in order to live an outrageously healthy life.

Yours in Health,

Dr. Greg S. Tomalin, D.C.

It is in the moments of decision that our destiny is shaped.
—Anthony Robbins

Chapter 1

How Your Body Works

There are three things I vividly remember from an embryology class I took in chiropractic school at Logan College. The first was the incredible complexity of the development of the human body; the second was why the health of the nervous system was so critical to normal development and health; and the third was the absence of my friend Bob (I always wondered how he passed that class).

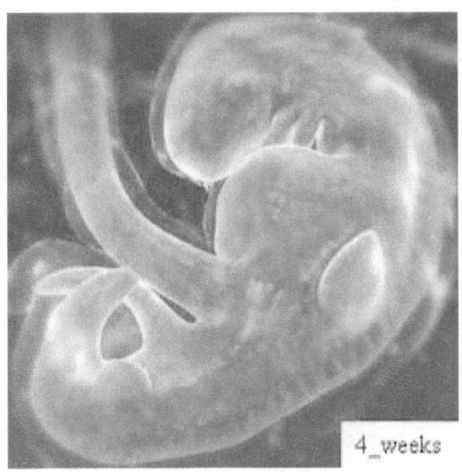

4_weeks

Starting at conception, the body begins to take shape, beginning with the development of a long cylindrical tube that forms the spinal cord. From this cord, little bumps begin to form that make up all the internal organs as well as the arms and legs. This process continues the way God intended for nine months until a fully formed human being emerges. As long as all the organs function in a tightly controlled balance, the child will continue to thrive and grow into a healthy adult.

While you sit calmly reading this, your body is extremely active. The trillions of cells that make up your body are busy performing thousands of delicately balanced processes that make your life possible. The brilliance with which your body controls this complex chemistry is truly divine. As long as your body is able to keep the process going, you remain healthy and vibrant. But if there is a disruption in any of the body's processes, the entire system loses its ability to perform correctly, and disease emerges. Disease is abnormal homeostatic cell function. This is fundamental to understanding your health, so let's talk about the process.

The body's ability to regulate and control the delicate balance of all life processes is called *homeostasis*. The term is derived from the Greek words for *same* and *steady*; it refers to the way the body acts to maintain a stable internal balance. For example, your body works to maintain a carefully regulated internal temperature of 98.6 degrees. If you go outside on a warm day and work, your body will begin to sweat in an effort to keep your temperature down. You may also begin to breathe more deeply in an effort to keep your tissues supplied with oxygen during a period of increased demand.

Disease and disability result when the body is stressed beyond its ability to maintain homeostasis. This stress can come from several sources, including interference with or irritation of the nervous system, which is the master controller of the body, due to poor diet, lack of exercise, or excessive emotional stress. One is no more important than another. In this chapter, you will learn about how the systems in your body maintain their health. A brief discussion of the digestive system, the cardiovascular system, and the immune system will be followed by an in-depth discussion of the largest system, the neuromusculoskeletal system—your nerves, muscles, and bones.

Digestive System

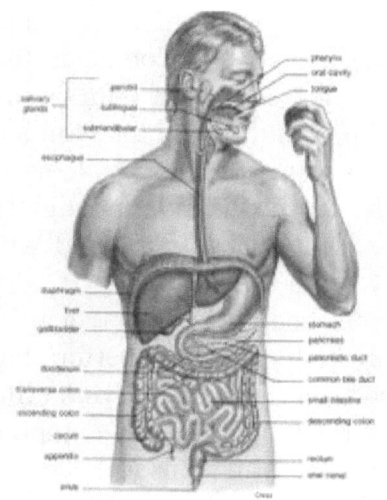

When we eat such things as bread, meat, and vegetables, they are not in a form that the body can use. Everything we eat and drink must be broken down into smaller molecules by means of digestion before it can be absorbed and used by the body.

The digestive system includes the digestive tract and its accessory organs, which process food into molecules that can be absorbed and utilized by the cells. Food is broken down, bit by bit, until the molecules are small enough to be absorbed, and the waste products are eliminated. The digestive tract, also called the gastrointestinal (GI) tract, includes the mouth, esophagus, stomach, small intestine, and large intestine.

As long as this system works correctly, the nutrients you consume can be extracted and absorbed into the bloodstream to be delivered to your individual cells. The digestive system is the means by which you get all the individual nutrients that are needed for health. Every single molecule in your body arrived there in the same way: at some point in the recent past you ate it. This is an important concept to remember, because if your diet does not include everything your body needs, you will lose some of the richness of your health. Consuming food is much like depositing money into your checking account. If you don't have enough resources there, you can't afford to keep fixing your house.

Cardiovascular System

The cardiovascular system, sometimes called the circulatory system, consists of the heart, which is a muscular pumping device, and a closed system of vessels called arteries, veins, and capillaries. The role of the cardiovascular system in maintaining homeostasis is to transport

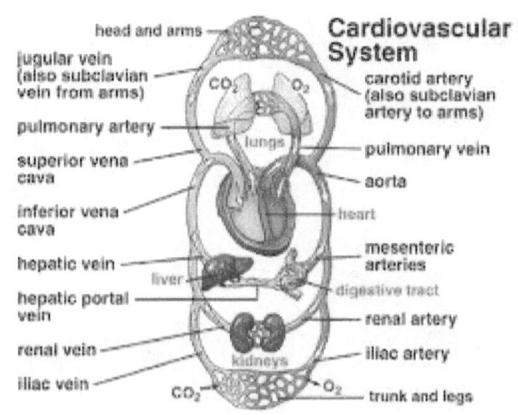

nutrients and oxygen from the digestive system to all the cells in the body and to transport waste and carbon dioxide to be eliminated.

As long as your cardiovascular system works correctly, your body has an enormous capacity to adapt to just about any external demand. For example, when you exercise, your heart pumps faster and your blood pressure increases in order to supply more oxygen and nutrients to your tissues. When you are cold, your blood vessels constrict in some areas of the body—the back of your arm, for instance—in an effort to conserve heat. When your body is injured, the blood vessels open up to allow white blood cells to enter the area to fight infection and speed healing.

Unfortunately, the cardiovascular system is the one that fails most often due to an unhealthy lifestyle. Heart disease is the number one cause of death in the United States, and it is also responsible for tragic disability in millions of Americans. But the vast majority of heart disease is completely avoidable by making some simple lifestyle changes, exercise being the most important. The key phrase when it comes to your cardiovascular health is *Use it or lose it*. If you don't get out and exercise and work your cardiovascular system, you will lose it. We'll talk more about this in the chapter on exercise.

Immune System

Studying the immune system is like watching a science fiction movie where nebulous blobs slither around surrounding and consuming unsuspecting prey. But unlike the blobs that threaten the existence of humankind, these blobs actually ensure our survival. They are the cells of the immune system. The immune system is a complex army of cells whose sole job is to protect the important cells in the body by eliminating the harmful ones.

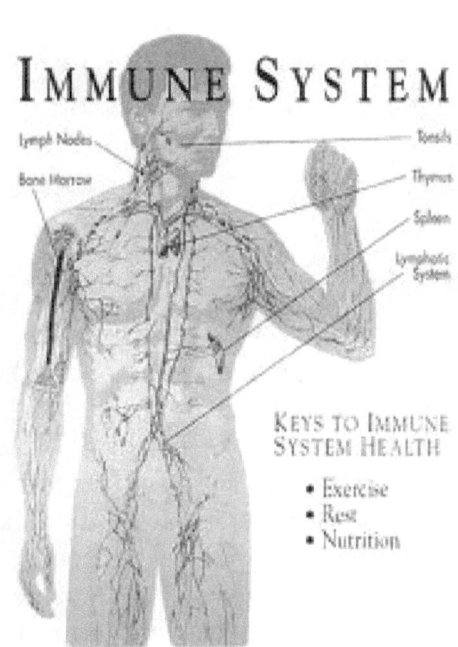

IMMUNE SYSTEM

Lymph Nodes
Bone Marrow
Tonsils
Thymus
Spleen
Lymphatic System

KEYS TO IMMUNE SYSTEM HEALTH

• Exercise
• Rest
• Nutrition

Every minute of the day, we are exposed to dangerous bacteria, viruses, and fungi. The cancerous cells that are always developing, if allowed to grow, will cause illness or death. In fact, the bacteria and viruses in and around our body outnumber our cells. They are everywhere. What keeps these invaders in check is the immune system.

When your immune system is compromised due to improper diet, stress, being subluxated (I'll explain this in the next chapter), or for any other reason, you lose some of your ability to eliminate the harmful cells and invaders in your body, resulting in disease. Just as with your body's other systems, your lifestyle choices have a profound impact on how well your immune system functions.

The Neuromusculoskeletal System

Bones and Joints

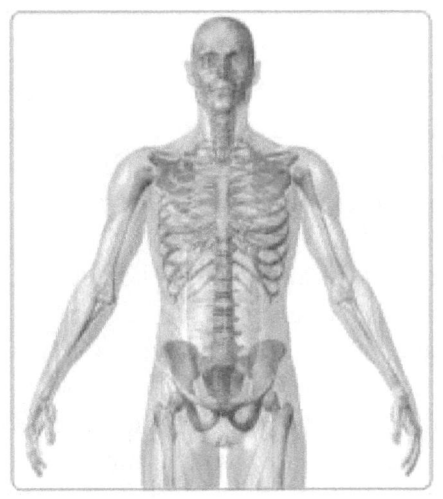

The human skeleton is made up of more than 200 bones, all connected by joints. Your bones are responsible for creating your body's general shape, and they protect your internal organs and manufacture blood cells. Each bone is made up of two compounds: a protein meshwork of collagen and a calcium salt called hydroxyapatite.

The collagen fibers that make up the basic structure of your bones give them a great deal of resilience and resistance to breaking when twisted, bent, or impacted. It is actually the loss of this collagen meshwork and not just a loss of calcium that is responsible for the bone weakness associated with conditions such as osteoporosis. The other component of bone, hydroxyapatite, is a crystalline calcium salt that is integrated into the collagen meshwork. Hydroxyapatite gives the bones rigidity and resistance to crushing under pressure.

Bones can be compared to steel-reinforced concrete, where the collagen meshwork acts like the steel meshwork in the concrete and the hydroxyapatite acts like the concrete that surrounds the steel. Together they form a tough, resilient, and rigid framework that supports the rest of the body. But because your bones are rigid and do not bend, you wouldn't be able to move if it weren't for your joints.

Joints are much more than simply a place where the ends of two bones meet. They are complicated systems of ligaments, tendons, membranes, and cartilage that allow the bones to move in a smooth, stable, and controlled way. Joints take different forms depending on their function and the particular stresses they have to endure. For example, the joints between your sternum (breastbone) and your ribs are simple joints consisting only of fibrous collagen. They are designed to be simple because the front part of your rib cage doesn't have to move very much in relation to your sternum. The shoulder joint, on the other hand, is extremely complex and requires a whole host of muscles, ligaments, and tendons all working in concert in order to move properly. If any one of the muscles or other structures of the shoulder are damaged, pain, instability, or loss of function may result.

Muscles

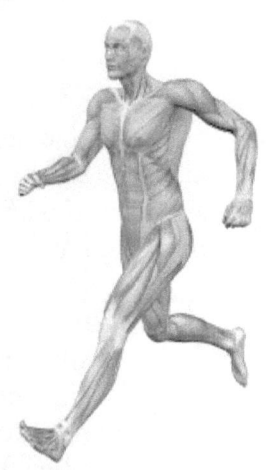

There are more than 650 muscles in your body, and they have one purpose: to create movement. While your bones give your body its framework, the muscles give your body motion. There are more than three times as many muscles in your body as there are bones, and each one fills a particular role in creating movement. Like bones, your muscles contain collagen for strength and resilience, but instead of calcium salts, muscles contain a specialized type of cell that has the unique ability to contract when stimulated by the nervous system.

There are three types of muscle: smooth muscle, cardiac muscle, and striated muscle (also called skeletal muscle). Smooth muscle surrounds the organs of the digestive tract and the arteries. In the digestive tract, smooth muscle moves the food through the digestive system, while the smooth muscle that surrounds the arteries regulates blood flow throughout the body. Unlike skeletal muscles, smooth muscles are involuntary, meaning that we do not have conscious control over them.

Cardiac muscle, as its name implies, is found only in the heart. What differentiates cardiac muscle from all other muscle is that it rhythmically contracts on its own, regardless of stimulation by the nervous system. As a matter of fact, if two independent cardiac cells, each rhythmically contracting to its own beat, are put in contact with each other, they will begin beating in unison. And it's a good thing, or our heart wouldn't beat regularly.

The third type is skeletal muscle. This is the type that we can consciously control and the type that is of most interest to us, because it is the muscle responsible for our posture and movement. Every skeletal muscle attaches to at least two different bones. As they contract, they draw the bones together, using the joints as hinges and allowing controlled movement to take place.

Take, for example, the elbow joint. Compared to some of the other joints, such as the shoulder or hip, the elbow is a relatively simple hinge. Yet, there are more than a dozen muscles that cross the elbow, all of which contribute to its normal movement. If any of these muscles do not fire in a coordinated fashion, or if some are tighter than they should be, or if some are weaker than they should be, abnormal joint function and pain will likely result.

Abnormal posture and joint motion resulting from muscle spasms or from weak or uncoordinated muscles is very common, especially in people who work at a desk all day. Because muscles become weaker if they are not exercised, it is important to include daily exercise as part of your routine.

The Nervous System

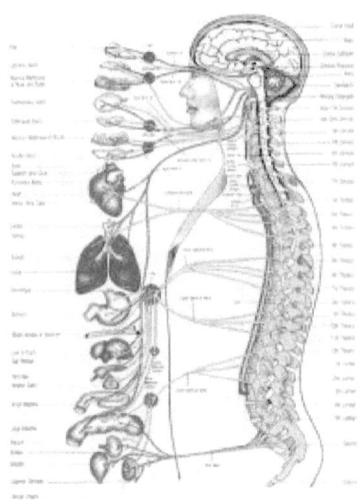

The nervous system is made up of trillions of highly specialized nerve cells, each of which communicates with hundreds or thousands of others through tiny electrical pulses. The nervous system is composed of two other systems: the central nervous system, your brain and spinal cord; and the peripheral nervous system, the nerves that run from your spine to the rest of your body. The nervous system is called the master controller, because it is responsible for the control and regulation of all body functions including our senses, movement, and balance.

There are three major types of nerves: pain nerves, motor nerves, and postural nerves, or more correctly, proprioceptors. Pain nerves do just what their name implies—they allow us to feel pain. Whenever something in our body hurts, it is because the pain nerves in the area are being stimulated and sending signals to the brain to create the sensation of pain.

Motor nerves are responsible for controlling our movement by stimulating muscles to contract. You are able to hold this book in your hands because the motor nerves are contracting the muscles in your hands and arms. If these nerves aren't able to function correctly, you can experience weakness or even paralysis in the muscles they control.

The third type of nerve is the proprioceptors, or in simple terms, postural nerves. They are responsible for sending information to the brain about where your body is and what it's doing. For example, if you close your eyes and hold your arm out to the side, you can tell exactly where your arm is even though you can't see it, because the postural nerves of the arm and upper back tell the brain where your arm is. Many people discover what happens when their postural nerves aren't working correctly when they have too much to drink. Alcohol partially disrupts your postural nerves, making it difficult to touch your finger to your nose when your eyes are closed or to walk a straight line with your eyes open.

In the next section, we will pull together this information on bones, joints, muscles, and nerves in a discussion of body mechanics.

The Four Pillars of Body Mechanics

The human body is an amazingly complex system of bones, joints, muscles, and nerves, designed to work together to accomplish one thing: movement. Movement is one of the defining characteristics that separate us, as animals, from plants, bacteria, and fungi. Everything about the human body is designed with movement in mind: nerve fibers stimulate muscles to contract, muscles contract to move bones, bones move around joints, and the nervous system controls it all.

Research has shown that movement is so critical to our body's health that a lack of movement has a detrimental effect on everything from digestion to our emotional state, immune function, ability to concentrate, how well we sleep, and even how long we live. *Subluxation* is a movement deficiency and alignment problem that is becoming epidemic (more on this later).

The bottom line is that if your lifestyle does not include enough movement, your body cannot function efficiently. Consequently, three things will happen:

- you will not be as physically healthy as you should be, and you will suffer from a wide variety of physical ailments from headaches to high blood pressure

- you will not be as productive as you should be because of reduced energy levels and an inability to focus mentally

- because you have less energy, your activity level will tend to drop off even further over time, creating a downward spiral of reduced energy and less activity.

This reduced functioning will take you to a point where even the demands of a sedentary job leave you physically exhausted by the end of the day.

Pillar One: Posture

The ancient Japanese art of growing Bonsai trees is fascinating. Bonsai trees are essentially normal shrubs that have been consistently stressed in a particular way for a long time to create a posture that is never found in nature. Depending on how the tree is stressed while it grows, it might look like a miniature version of a full-sized tree, or it might look like a wild tangle of branches with twists and loops. Every day in my practice, I see the human equivalent of Bonsai trees walk through my door—people with an unnatural posture due to continual stress on their bodies.

The most immediate problem with poor posture is that it creates chronic muscle tension, because the weight of the head has to be supported by the muscles instead of the bones. This is referred to in the literature as Forward Head Posture (FHP). This effect becomes more pronounced the farther your posture deviates from your structural center (see illustration). The Mayo Clinic has said that FHP leads to long-term muscle strain, disc herniations, arthritis, and pinched nerves.

Think about carrying a briefcase. If you had to carry your briefcase with your arms outstretched in front of you, the muscles of your shoulders soon would be completely exhausted, because carrying the briefcase far away from your structural center  places undue stress on your shoulder muscles. If you hold the same briefcase down at your side, your muscles do not fatigue as quickly because the briefcase is closer to your structural center and the weight is supported by the bones of the skeleton rather than the muscles.

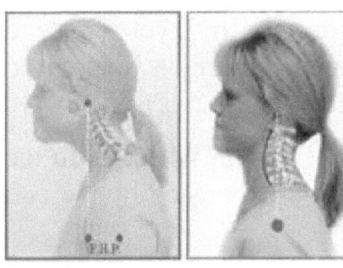

You can imagine the amount of pressure on your muscles, ligaments, and spinal structures when your head (about the weight of a bowling ball) sits forward of your shoulders as you read, sit, and work on the computer day after day. According to Renee Calliet, M.D., "as the head, weighing ten pounds, sits over your structural center, the load is only ten pounds. However, for every inch forward your head is

over your center, the weight increases by ten pounds." Therefore, if your head is three inches over your shoulders, your muscles and spinal tissues are holding thirty pounds! This eventually leads to *subluxation*, a loss of cervical curve, and the health problems it causes.

In some parts of the world, women carry big pots of water from distant water sources back to their homes. They can do this without significant effort because they balance them on their heads, carrying them at their structural center and allowing the strength of their skeleton to bear the weight rather than their muscles.

Correcting bad posture and the physical problems that result is accomplished by doing two things. The first is to eliminate as much "bad" stress from your body as possible. Bad stress includes all the factors, habits, and stressors that cause your body to deviate from your structural center. This can include a poorly aligned chair and desk at work, the wrong seat adjustment in your car, or carrying too much weight in a purse, briefcase, or backpack. The second is to apply "good" stress to the body in order to move your posture back toward your structural center. Centering your body by improving your posture is critically important to improving how you

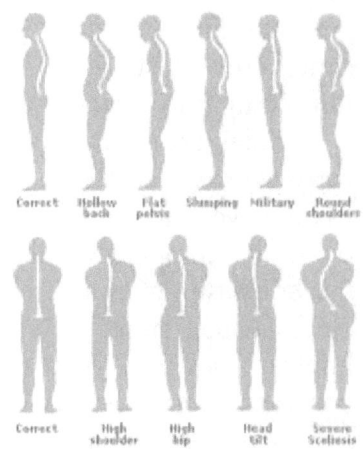

feel. This is accomplished through a series of exercises, stretches, and changes to your physical environment that will help correct your posture.

Pillar Two: Mobility

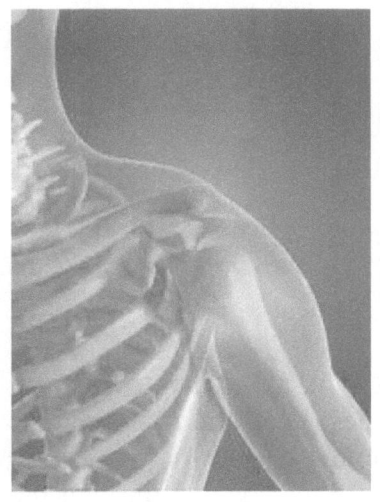

Imagine waking up one morning with a frozen shoulder that prevents you from moving your upper arm more than a few inches in any direction. How much would that affect your ability to do your job? How much would that affect your ability to drive your car or even to dress yourself? How much would that affect your ability to concentrate on anything other than your shoulder? Obviously, if your shoulder doesn't move correctly, it will have a dramatic impact on your life. The same is true with mobility in every part of your body. If things aren't moving the way they should, it will be difficult or even impossible to function at work or to take care of the demands of everyday life. You will have difficulty thinking clearly or focusing on a task.

Over the years, I have had a number of patients come into my office with severe low back pain that came on suddenly when they did something as simple as bend down to pet their dog, put on their socks, or pick up the newspaper. Just about everyone would agree that a person's body should be able to handle something as simple as bending over to pick up a newspaper or put on socks, right? So what happened?

In every one of these cases, we find that many of the joints in their body are barely moving at all: they are locked up, or *fixated*. When the joints in one area of the body do not move as they should, other areas are forced to move more than they were designed to in an effort to compensate. This creates significant stress on the areas that pick up the slack, which soon leads to pain and inflammation. At the same time, the areas that don't have normal movement will slowly worsen as the muscles continue to tighten, the joints stick together, and the ligaments and tendons shorten. This leaves the body in a very unstable condition. Left unchecked, this process will continue until the body can hardly move at all, and the person suffers flare-ups of pain at the slightest provocation.

Most of us have seen people who have lost most of their normal mobility. They look like their whole body is frozen stiff. This is especially prevalent among the elderly. Contrary to popular belief, this is not the effect of aging but the inevitable effect of not maintaining the body's mobility through exercise, stretching, and chiropractic care. There are a lot of people in their sixties or seventies or older who are stronger and more flexible than the average person in their thirties, simply because they 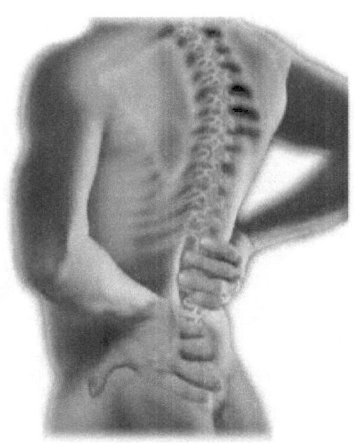 keep on exercising. Maintaining mobility is critical in order to live free from pain and disability.

Just as in developing good posture, it is necessary to perform exercises and stretches to keep your muscles, ligaments, and tendons flexible and healthy. In addition, all the joints in your body must be kept moving correctly. This can be achieved to a degree through the exercises and stretches in this book, but most people also find routine chiropractic adjustments to be very beneficial.

Pillar Three: Strength

 Strong muscles keep your body upright and allow you to move. Good muscle strength and balance are critical to minimize muscle tension and maintain proper posture. Your muscles function much like the wires that hold up a tall radio or television antenna. If the wires are equally strong on all sides, the antenna will stand up straight. If one of the wires becomes weak or breaks, the antenna will either lean to the side or collapse. The same is true with your body. If the muscles on all sides of your spine are balanced and strong, your body will stand up straight and strong. Unfortunately, most people don't have strong, balanced muscles.

Muscles get stronger or weaker in response to the demands placed on them. Since most of us sit at a desk, drive a car, and sit on the sofa at home, some muscles are not challenged and consequently become weak. At the same time, the muscles that are constantly used throughout the day become strong. This imbalance of muscle strength contributes to poor posture and chronic muscle tension. Left unchecked, muscle imbalances tend to get worse, not better, because of a phenomenon called *reciprocal inhibition*.

Reciprocal inhibition literally means *shutting down the opposite*. Simply put, for all the muscles that move your body in one direction, there are opposing muscles that move the body in the opposite direction. In order to keep these muscles from working against each other, when the body contracts one muscle group, it forces the opposing group to relax—it shuts down the opposite muscles.

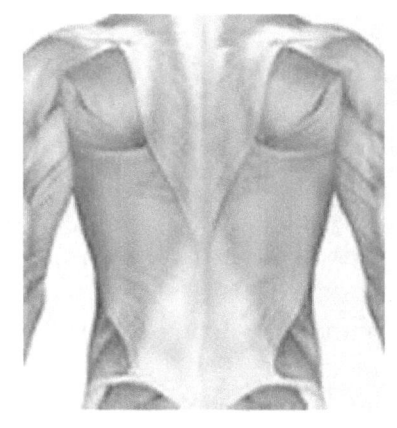

This phenomenon is especially important to people who work at a desk, because all day the same muscles in the upper back and chest are used and the body shuts down the opposite muscles in the middle back. Over time, the middle back becomes weak, because the muscles are not being worked as are the muscles in the front of the body. This contributes to poor posture, chronic muscle spasms, and pain.

The easiest way to correct this imbalance is to do specific exercises that will increase the strength of the back muscles and to get chiropractic care. Once the muscles in your middle back are strong, the tightness and poor posture simply disappear.

Pillar Four: Coordination

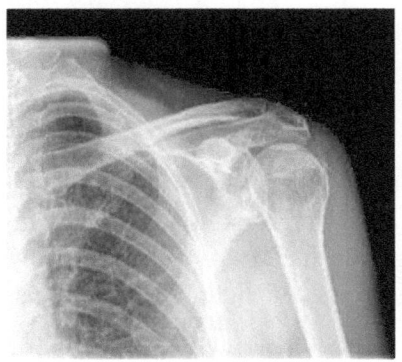

I was playing hockey (the Canadian government requires this to keep my citizenship while living in the United States) when I began suffering from shoulder pain. I took x-rays, ordered an MRI, did all the orthopedic tests in an attempt to figure out what was wrong with my shoulder, but everything turned up normal. I was young and healthy, and I had good strength and great flexibility. There was no specific injury to the shoulder. Since the shoulder is a mobile and unstable joint, we know that if all its muscles are not contracting in the correct order or with the right amount of tension, the result is increased mechanical stress of the shoulder joint, resulting in pain.

I finally did a series of very simple, lightweight exercises on a daily basis in order to reestablish normal shoulder coordination. The results were immediate and profound. Not only did my pain completely disappear, but my strength and range of motion improved as well. It turned out that my only problem was that my muscles were not coordinated correctly. Although posture, joint mobility, and muscle strength are all important, they are not the whole story. We also must have coordinated control over our muscles and joints if we want to enjoy good body mechanics.

Healthy coordination is simply the result of using the body in the manner in which it was designed. Exercises such as walking, swimming, rock climbing, yoga, Pilates, bicycling, martial arts, and body building all help to improve muscle coordination, whereas working at a desk, reading, and watching television do the opposite. Without realizing it, most people are in a dramatic state of muscle imbalance. This occurs simply because they sit for many hours every day and do not perform exercises that work to keep their muscles properly coordinated. This muscular imbalance contributes to muscle tightness, restricted movement, and joint pain.

The Mechanics of Your Spine

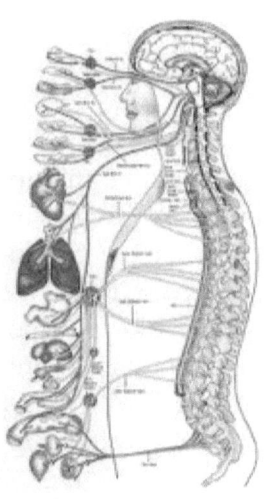

Now that you're an expert on body mechanics and you understand why it's so important that the skeletal system, muscular system, and nervous system work together in a tightly coordinated way, let's take a look at the single most complex and important system of bones, muscles, and nerves in your body—your spine.

The spine is one of the most complex systems in the body, consisting of nearly a hundred intricate joints and trillions of nerve pathways connected by a complicated meshwork of ligaments, tendons, cartilage, and muscles. The spine is designed to do three things simultaneously.

- protect the spinal cord that serves as the primary communication and lifeline between your brain and the rest of your body

- serve as a structural support upon which all your organs and upper body rest

- provide mobility and flexibility, allowing you to bend forward to touch your toes or pick up your child to play hockey or throw a baseball, or simply to turn your head.

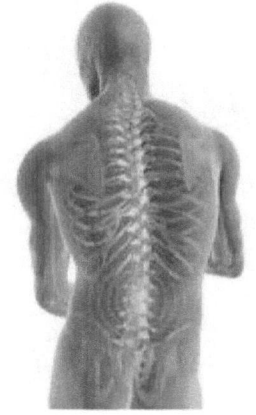

However, because the body is mobile and flexible, it is also unstable and susceptible to injury.

In order to function correctly, all the bones, joints, muscles, and nerves have to work in perfect coordination to maintain your posture, strength, and movement. A disruption in the position or movement in any of the bones of the spine, or a loss of muscle balance or coordination, will impose a significant stress on the spine.

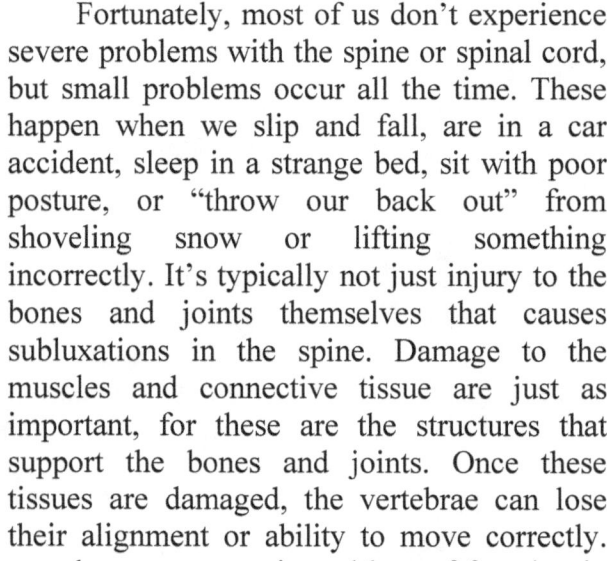

Fortunately, most of us don't experience severe problems with the spine or spinal cord, but small problems occur all the time. These happen when we slip and fall, are in a car accident, sleep in a strange bed, sit with poor posture, or "throw our back out" from shoveling snow or lifting something incorrectly. It's typically not just injury to the bones and joints themselves that causes subluxations in the spine. Damage to the muscles and connective tissue are just as important, for these are the structures that support the bones and joints. Once these tissues are damaged, the vertebrae can lose their alignment or ability to move correctly. When this happens, it not only can cause pain and loss of function in the back, but it can affect other areas of the body as well.

The spine is made up of a stacked set of bones called the *vertebrae.* These are like the bricks upon which our entire structure is built. Each vertebra consists of a *vertebral body*, which is a large oval-shaped solid block of bone, and a *vertebral arch*, which is located on the back of the vertebral body and creates the space through which the spinal cord runs.

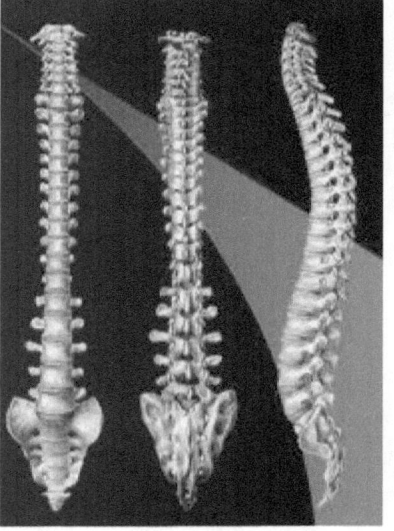

Each vertebra is attached to two adjacent spinal vertebrae with a disc between them. These discs, technically called *intravertebral discs*, are thick pads of fibrocartilage that act as shock absorbers and give the spine its ability to flex and twist. The disc itself is kind of like a jelly-filled donut. It has an outer fibrous portion called the *annulus* and a soft center called the *nucleus pulposis*. A disc herniation occurs when the fibers in one portion of the annulus are torn, allowing the nucleus pulposis to push through the annulus. This creates a bulge in the outer disc, much like a tire that develops a bulge when there is a break in one of the underlying supporting fibers.

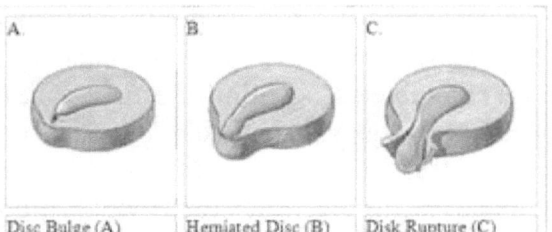

Disc Bulge (A)   Herniated Disc (B)   Disk Rupture (C)

Disc bulges do not always lead to pain, but quite often they do. The pain may come from the irritation of the nerves within the disc itself, or it may be caused by the disc bulge impinging upon a nerve that runs through the area. As we age, our discs tend to dehydrate and break down. This process is accelerated by smoking and not keeping yourself fully hydrated, as well as by a sedentary lifestyle.

Between each pair of vertebrae and behind the disc, there is a small space where the nerves exit from the spinal cord and run to the rest of the body, called the *vertebral foramen.* Foramen is an impressive medical term that simply means *hole.* Vertebral foramina can become compressed when a disc bulge presses into the area, when inflammation causes the tissues in the area to swell, or when the intravertebral disc becomes dehydrated. This can cause excruciating pain that can radiate to other areas of the body.

Ligaments bind the vertebrae together, and tendons attach muscles to each segment. The ligaments and tendons help absorb shock and restrict the amount of movement between the spinal vertebrae. Unfortunately, these ligaments and tendons can be damaged whenever spinal vertebrae are forcefully moved beyond their normal limits, such as in a whiplash or a sports injury. An injury to a ligament is a *sprain*; an injury to a tendon or muscle is a *strain*.

Muscles attach to the bony extensions of the vertebrae and provide movement in the spine by contracting in a highly coordinated way. Like ligaments, muscles absorb shock and release it in a controlled way. When your heel strikes the ground as you walk, it's your muscles that dissipate the shock before it reaches your head so that your teeth don't clatter together with each step.

The spine forms the protective housing for the spinal cord, which begins at the brain stem at the base of the skull and extends like a wire down the length of the spine. The spinal cord sends out nerve branches that send and receive signals from every cell in the body. The close relationship between the spine and the spinal cord means that damage to a disc or vertebra can affect the spinal cord or the nerves associated with it, causing pain or abnormal function in the structures controlled by those nerves.

The spine is divided into four regions. (Picture 25)

The upper seven vertebrae in the neck are collectively called the *cervical spine*, with the skull sitting directly upon the first cervical vertebra, also called the *Atlas*. The middle twelve vertebrae are called the *thoracic vertebrae*. Each thoracic vertebra has a pair of ribs attached to it. Taken together, the twelve pairs of ribs protect several of your internal organs and are critical for breathing. The lower five vertebrae are referred to as the *lumbar vertebrae*. Because they bear the full weight of your upper body, they are the most frequently injured. The lowest region of the spine is called the *sacrum*. In young children the sacrum is made up of five vertebrae, just like the lumbar spine. Later in childhood, these five vertebrae fuse together to make one solid bone called the sacrum.

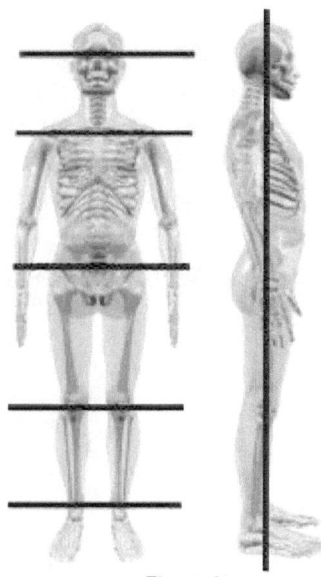

Figure 1

When viewed from the front or back, the spine should appear perfectly straight and symmetrical, reflecting the fact that your body is symmetrical. When viewed from the side, however, four major curves should be seen, one in each of the cervical, thoracic, lumbar, and sacral regions. In the cervical and lumbar regions of the spine, the curves bend backward in *lordotic* curves. The curves in the thoracic and sacral regions bend forward in *kyphotic* curves. As strange as it may seem, these curves add a considerable amount of strength and resiliency to the spine. Think of the curves as springs that allow the spine to flex and absorb shock better than if it were straight. In fact, when a region of the spine loses its normal curve, as often happens in the neck following a whiplash injury, the discs that separate the vertebrae begin to degenerate.

Because of the complexity and instability of the spine and its potential for affecting so many systems in the body, chiropractors work on this structure more than any other. Problems in the spine can come from a variety of sources:

- Discs can become herniated and compress nerves to the legs or arms

- Joints between the vertebrae can become stuck

- Bones, ligaments, or joints can be injured

- Vertebrae can become subluxated or misaligned

- The muscles surrounding the spine can be injured

- Muscle spasms can develop due to overuse or injury

- Inflammation from overuse, injury, or disease can irritate the spine

Each of these problems needs to be identified and properly treated if you are to enjoy the optimal health of which your body is capable. Just cracking your back is not enough.

Review

Let's review quickly before moving on. We discussed the major systems that are responsible for our health: the digestive system, the cardiovascular system, the immune system, and the neuromusculoskeletal system.

We discussed the four components of the neuromusculoskeletal system: the bones and joints that serve as the framework for the body, the muscles that are responsible for movement, and the nervous system that controls it all. We discussed the four pillars of body mechanics: posture, mobility, strength, and coordination.

You learned that when your posture deviates from your structural center, your muscles tighten up, resulting in pain and decreased mobility. You learned that once you begin to lose mobility, it usually worsens unless you perform specific adjustments, exercises, and stretches to reestablish normal motion. You learned that muscles will become either stronger or weaker depending on how much they are used and that muscular imbalances can lead to chronic muscle spasms and pain. You learned about the important role of coordination in achieving good body mechanics. Finally, you learned that the main connection from the brain to the rest of the body is the spinal cord. Your spine is an incredibly complex structure for your overall health, due to its close relationship to the spinal cord and nerves. If the spine is healthy, then the central nervous system can effectively communicate and coordinate all the body's functions. But if there is any misalignment, impingement, or interference with this communication, then pain and lack of function result. You learned that in order to have vibrant health, you have to make lifestyle choices that won't undermine your body's ability to be healthy.

Chapter 2

Chiropractic Care

The word *chiropractic* comes from the Greek words *cheir* (hand) and *praxis* (action). It literally means "done by hand." Instead of prescribing drugs or performing surgeries, chiropractors use manual adjustments of the spine and joints, exercise therapy, nutritional counseling, and instruction in lifestyle changes to allow the body's natural state of health to fully express itself.

Like medicine, chiropractic is based upon scientific principles of diagnosis through testing and empirical observation and treatment based on the doctor's rigorous training and clinical experience. But unlike conventional medicine, which focuses on attempting to treat disease once it occurs, chiropractic attempts to improve the health of the individual in an effort to avoid illness in the first place. Most people would rather be healthy and avoid illness if they can. This is one of the main reasons for the upsurge in the popularity of chiropractic. People are recognizing the benefit of seeking an alternative to traditional medicine—one that will help them ach and maintain optimal health.

Corrective care/wellness chiropractors understand that much of our pain, sickness, and disease is due to subluxation, or the misalignment and abnormal motion of the vertebrae in the spinal column.

Chiropractic works by removing these subluxations in the spine to relieve pressure and irritation on the nerves, restoring joint mobility and returning the body to a state of normal function. The lack of motion in spinal bones from subluxation is a deficiency in brain nutrients. The brain requires *proprioceptive* (movement) input from the body. This is not controversial in the medical community, but few medical doctors realize that subluxation is a movement deficiency problem.

Numerous studies have demonstrated that chiropractic corrective care is one of the most effective treatments for back pain, neck pain, headaches, whiplash, sports injuries, and many other health problems because it improves posture and motion of spinal joints. It has even been shown to be effective in reducing high blood pressure, decreasing the frequency of childhood ear infections, and improving the symptoms of asthma.

> Posture affects and moderates every physiologic function from breathing to hormonal production. Spinal pain, headache, mood, blood pressure, and lung capacity are among the functions most easily influenced by posture. The corollary of these observations is that many symptoms, including pain, may be moderated or eliminated by improved posture.
>   —*AJPM* 1994, N. Shealy M.D., R. Cady M.D.

The chiropractic approach to health care is holistic, meaning that it addresses your overall health. Because many lifestyle factors such as exercise, diet, stress, proper rest, and your environment impact your health, I recommend changes in lifestyle—eating, exercise, stress management, and sleeping habits—in addition to chiropractic corrective care.

The History of Chiropractic

Manual manipulation of the spine and other joints in the body has been around for a long time. Ancient writings from China and Greece dating between 2700 B.C. and 1500 B.C. mention spinal manipulation and the maneuvering of the lower extremities to ease low back pain. Hippocrates, the famous Greek physician who lived from 460 to 357 B.C., published a text detailing the importance of manual manipulation. In one of his writings he declares, "Get knowledge of the spine, for this is the requisite for many diseases." Evidence of manual manipulation of the body has been found among the ancient civilizations of Egypt, Babylon, Syria, and Japan, and among the Incas, Mayans, and Native Americans.

The official beginning of the chiropractic profession dates back to 1895, when Daniel David Palmer restored the hearing of Harvey Lillard by manually adjusting his neck. Dr. Palmer knew that he had discovered something incredible. Two years later, in 1897, he went on to establish the Palmer School of Chiropractic in Davenport, Iowa, which continues to train doctors of chiropractic to this day.

Throughout the twentieth century, chiropractic has gained considerable recognition and scientific support. Research studies have clearly demonstrated the value of chiropractic care in reducing health care costs, improving recovery rates, and increasing patient satisfaction. One large study conducted in Canada, the 1993 Manga Study, concluded that chiropractic care would save hundreds of millions of dollars annually in work disability payments and direct health care costs. Several major studies conducted by the U.S. government, the Rand Corporation, and others have all demonstrated the value of chiropractic care. Unfortunately, there are still many people who have never been to a chiropractor and don't understand what we do. My hope is that this book will help educate people about chiropractic corrective care and its role in the wellness revolution.

The Vertebral Subluxation Complex

What differentiates doctors of chiropractic from other health professionals is that chiropractors are the only ones who are trained to diagnose and treat spinal subluxations. The word *subluxation* comes from the Latin words *luxare*, meaning *to dislocate*, and *sub*, meaning *somewhat* or *slightly*. So the term *vertebral subluxation* literally means a slight dislocation or misalignment of the bones in the spine.

This term was adequate in the 1800s, when much was still misunderstood about the human body, but today subluxation embodies the complex of neurological, structural, and functional changes that occur when a bone is out of place. For this reason, chiropractors usually refer to subluxations of the spine as the Vertebral Subluxation Complex (VSC). There are five components that contribute to the VSC.

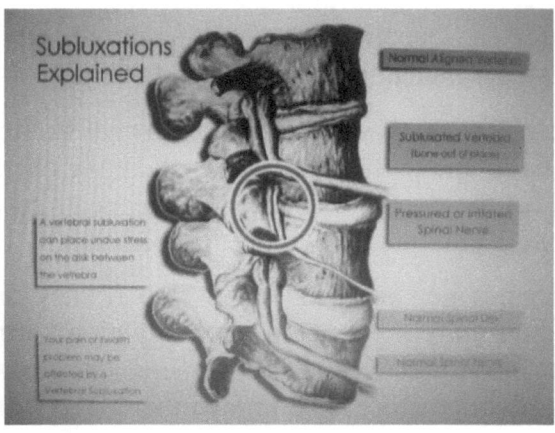

- The *bone component*, where the vertebra is out of position, not moving properly, or undergoing degeneration. This frequently leads to a narrowing of the spaces between the bones through which the nerves pass, often resulting in irritation or impingement of the nerve.

- The *nervous component* is the disruption of the normal flow of energy along the nerve fibers, causing the messages traveling along the nerves to become distorted due to a lack of spinal motion. The result is that all the tissues that are fed by those nerves receive distorted signals from the brain and consequently are not able to function normally. Over time, this can lead to a number of conditions, such as peptic ulcers, constipation, and other organ system dysfunctions.

- The *muscular component*. Since nerves control the muscles that help hold the vertebrae in place, muscles are an integral part of the VSC. In fact, muscles both affect and are affected by it. A subluxation can irritate a nerve, the irritated nerve can cause a muscle to spasm, the spasmed muscle pulls the attached vertebrae farther out of place, the nerve is further irritated, and a vicious cycle is set in motion. It's no wonder that very few subluxations go away by themselves.

- The *soft tissue component*. The VSC will also affect the surrounding tendons, ligaments, blood supply, and other tissues as the misaligned vertebrae tug and squeeze the connective tissue with tremendous force. Over time, the soft tissues can become stretched or scarred, leaving the spine with a permanent instability or restriction.

- The *chemical component* is the change in the chemistry of the body due to the VSC. Most often, the chemical changes, such as the release of a class of chemicals called *kinins* that increase inflammation in the affected area.

These changes get progressively worse if they are not treated correctly, leading to chronic pain, inflammation, arthritis, muscle trigger points, bone spurs, loss of movement, muscle weakness, and muscle spasm. Chiropractors have known the dangers of the vertebral subluxation complex since the birth of the profession. More and more scientific research is demonstrating the detrimental impact that subluxation has on the tissues of the body.

In order to be truly healthy, it is vital that your nervous system function free of interference from subluxations. Chiropractors are the only health professionals trained in the detection, location, and correction of the vertebral subluxation complex.

Chiropractic Treatment

Spinal adjustments to correct subluxations make doctors of chiropractic unique among health care professionals. An adjustment is the specific manipulation chiropractors apply to vertebrae that have abnormal movement patterns or otherwise fail to function normally. The objective of the chiropractic adjustment is to reduce the subluxation, resulting in increased range of motion, reduced nerve irritability, and improved function.

In *The 14 Foundational Premises for the Scientific and Philosophical Validation of the Chiropractic Wellness Paradigm* (2003), James L. Chestnut states that

> 'Chiropractic adjustments restore proper motion and concomitantly create bridges across unpotentiated (or depotentiated) synapses. The adjustment, and subsequent increased joint motion, stimulate the neuroplastic phenomenon of synaptogenesis and build pathways for delivery of the most important nutrient for heath and homeostasis ever identified: mechanoreception.'

The chiropractic adjustment is a quick thrust applied to a vertebra for the purpose of correcting its position, its motion, or both. Adjustments may be accompanied by an audible release of gas, making a cracking sound. This may be startling at first, but it is only the sound of nitrogen gas surging into the joint space. Occasionally minor discomfort is experienced, especially if the surrounding muscles are in spasm or if the patient tenses up during the adjustment. Sometimes cracking does not occur. This doesn't indicate a good or bad adjustment; it is simply due to the chiropractic technique or to muscle tightness.

There are two different types of chiropractors, with different philosophies and techniques. Like an M.D., the traditional chiropractor focuses on pain relief. In addition to manipulations, these practitioners use electrotherapies such as ultrasound, electrostimulation, diathermy, TENS, or lasers. Some of these chiropractors are trying to expand the scope of their practice to prescribe drugs. As in traditional medicine, as soon as your pain begins to subside, you are released from care. Often an underlying problem still exists, the spine is left to deteriorate, and future problems are certain. Little to no consideration is given to the patient's underlying problems, toxicities, deficiencies, or wellness. There are still such practitioners, but this type of chiropractic is becoming antiquated.

Corrective care/wellness chiropractic is the wave of the future, particularly because it is so much more than simply a means of relieving pain. Relief of pain and symptoms is important and usually happens very quickly, but it is not the primary focus. Ultimately, the goal is to restore the body to its natural state of optimal health and return the spine to its proper position and motion so that the body can heal itself.

I use a manual adjusting technique called Chiropractic Biophysics (CBP). This technique is one of the few that can remodel the position and motion of the spine over time. I also recommend postural exercises, nutrition management workshops, exercise rehabilitation, a massage therapist, and counseling on lifestyle issues. Since the body has a remarkable ability to heal itself and to maintain its own health, my primary focus is to remove those things that interfere with healing.

Wellness chiropractors understand that within each of us is an innate wisdom, a health energy, that will express itself as perfect health and well-being if we simply allow it to do so. Therefore, the focus of chiropractic care is to remove physiological blocks to the proper expression of the body's innate wisdom. Once subluxations are removed, health is the natural consequence.

Phases of Spinal Degeneration

Normal

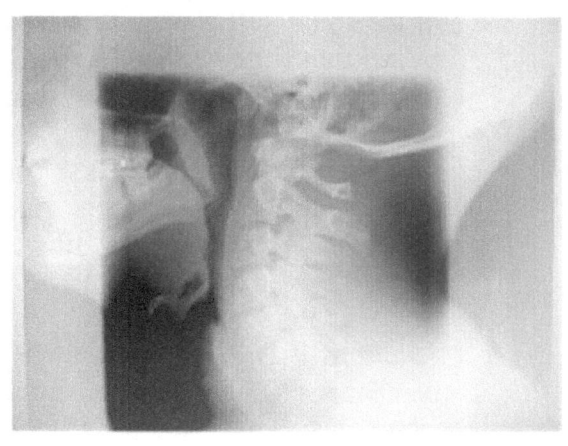

This is an x-ray of the side view of the cervical spine—the neck—of a person facing left. The 60-degree neck curve shown is called the *Arc of Life*, because it allows normal flow of messages from the brain down the spinal cord. The Arc of Life allows messages to be sent without any pressure or tension on the spinal cord. Notice that there is even spacing between the discs and that the vertebrae are squared off.

Phase One

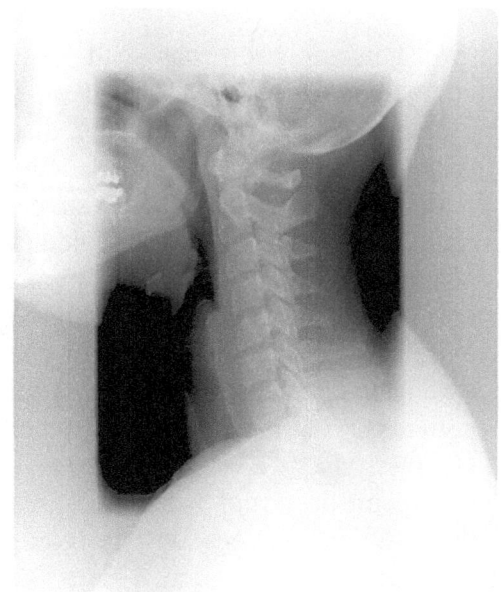

This phase of subluxation degeneration is characterized by a change in the cervical curve. Disc spaces have begun to narrow, and the vertebrae are beginning to change shape. The loss of the normal neck cure is adding stress to the spinal cord and spinal tissues. Segmental motion is becoming restricted, although overall motion may be normal. A serious problem is developing, usually without symptoms. More than 75 percent of cases show no symptoms or pain in Phase One. Chiropractic corrective care at this stage can range from 6 to 18 months to remodel the spine.

Phase Two

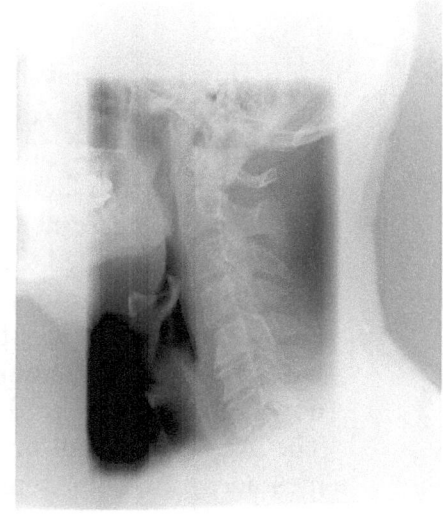

Phase Two is Degenerative Joint Disease (DJD), or arthritis of the spine. DJD is characterized by a loss of the cervical curve, a significant narrowing of the disc space, and calcium changes. X-rays show calcium spurs, or *osteophytes*, forming on the vertebrae. This condition affects nerve function and causes loss of range of motion in the neck. As severe as it is, however, only 50 percent of DJD patients have symptoms or pain. If the condition is not caught and corrected, it will certainly progress. Chiropractic corrective care in Phase Two can range from 1.5 to 2.5 years.

Phase Three

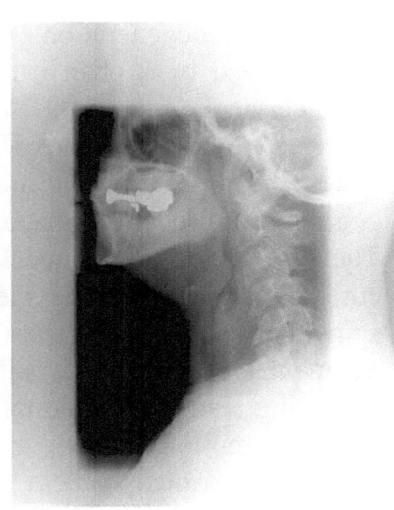

This phase has all the characteristics of DJD, only worse. The curve is nonexistent, the disc spaces are almost fused, the vertebrae have changed shape, arthritis is advanced, and calcium changes and nerve system problems are noted. This condition can be extremely serious, depending on the extent of nerve damage. Almost all patients will have pain or other symptoms. Even extensive treatment does not guarantee that the calcium changes will be gone, but adjusting the spine slows or stops further degeneration. This is the last phase before spinal fusion, so this is the last point at which spinal correction can occur.

Phase Four

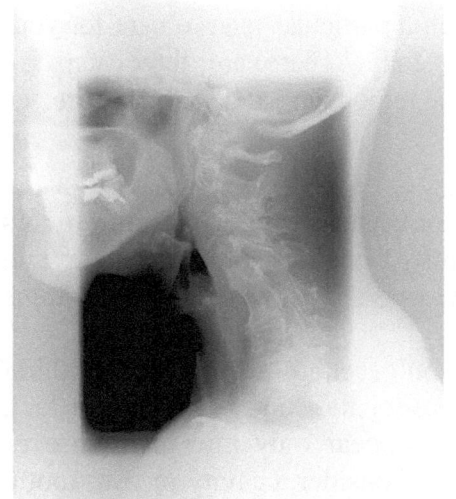

Phase Four is a serious condition in which neurological damage can limit the patient's health status and quality of life. The vertebrae have now fused together, and range of motion is extremely limited. These patients have severe structural and neurological problems. Although correction is impossible at this stage, many patients report improvement in their symptoms.

Spinal Surgery

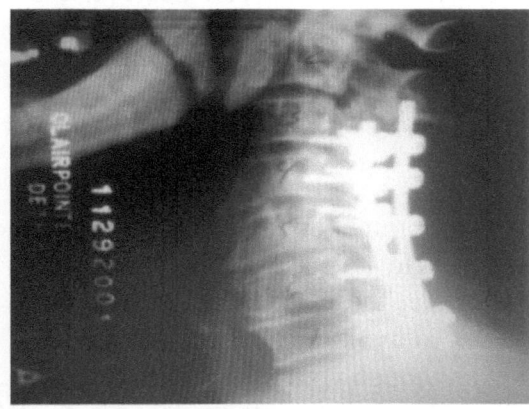

This is a patient in Phase Four who had surgery on his spine, the medical approach intended to restore the curve to the neck. Notice the 60-degree curved metal rod and screws the surgeon used to take pressure off the spine and try to restore the Arc of Life. Orthopedic surgeons know the importance of the cervical arc, but this patient is still on medications for his symptoms, because there was so much permanent damage to his nervous system that could have been prevented by correcting the subluxation in the first place.

The Three-Step Process in Chiropractic Corrective Care

My children's Grandpa Paul was a contractor who built houses. Chiropractic care is like building a house: certain things have to happen in a particular order for everything to stand strong and work as it should. If you try to put up your walls before you have a solid foundation, they will be weak and will eventually collapse. If you try to put on your roof before the walls are ready, you will run into the same problem. The same is true for your body. You have to go through a particular plan of care for your body to repair itself correctly and fully. There are three steps in chiropractic care: corrective care, active remodeling and wellness care.

Step 1. Corrective Care

Many people go to a chiropractor because they are in pain. In this first phase of care, the goal is to identify toxicities and deficiencies and begin restoring cell function to homeostasis. As this occurs, symptoms often subside.

Most people think that if they don't feel any pain there is nothing wrong with them, but pain is a very poor indicator of health. In fact, pain and other symptoms frequently appear only after a disease or other condition has become advanced. Consider a cavity in your tooth. Does it hurt when it first develops or after it has become serious? How about heart disease? It is well-known that the first symptom of heart disease is often death. In conditions such as cancer, heart disease, diabetes, stress, and problems with the spine, pain usually appears only in later stages. When you begin chiropractic care, pain may disappear, but the underlying condition remains and must be addressed.

Step 2. Active Remodeling

Most chiropractors regard the elimination of symptoms as the easiest part of a course of treatment. However, if all the chiropractor does is relieve the pain and stop there, the chances of the condition recurring are considerable. In order to avoid a rapid recurrence, it is necessary to continue treatment even though your symptoms are gone.

During the *active remodeling* phase, treatments are less frequent, but more work is done between visits. Depending on your circumstances, you might begin exercises, stretches, and traction at home to accelerate your spinal remodeling and healing. There may be mild flare-ups in your symptoms. This is normal. Flare-ups are bound to occur during active remodeling because your body has not fully recovered. Depending on the severity of your injury or condition and how long you have been suffering from it, this phase of your care may last anywhere from a few months to two years. When you're treated by a corrective care/wellness chiropractor, not only will you feel improvement and function, better but you also can see differences in your x-rays before and after treatment.

This patient was seeing a traditional back pain chiropractor once a month for two years before being referred to my office for chronic headaches, sinus problems, and ringing in his ears. When I asked why he was going once a month to be adjusted, his answer was frightening. It wasn't what he wanted, it wasn't what his chiropractor wanted, but his HMO covered twelve visits. I took x-rays because his first chiropractor hadn't. When he saw his x-rays he was angry, to say the least. "Dr. Greg, how could my spine look this bad when I've been going to a chiropractor once a month?" I explained the difference between relief care and corrective care. "Mark, if you work out once a month, will you see changes in your physique? No. If you diet once a month, will you lose weight? No. What you were doing was maintaining a bad spine. You never actually corrected it to its proper position." Mark finally got it and decided to follow the corrective care program. Here are his results after thirteen months.

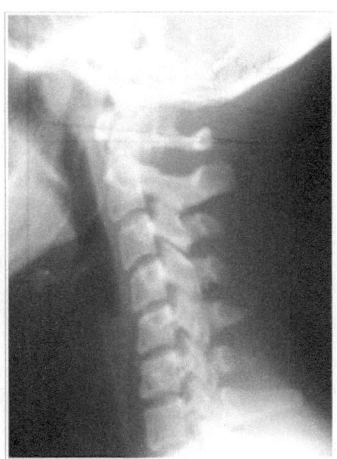 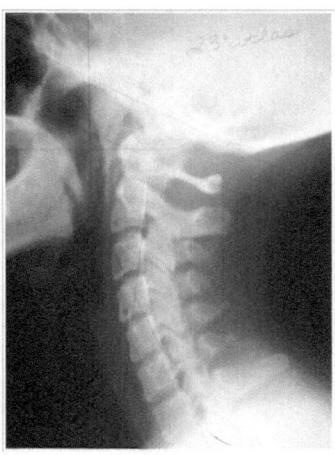

No more headaches, sinus problems, or ringing in his ears.

Before John saw me, his immune system was severely depressed. Any time someone would get sick around him, he would get sick. His nerve system wasn't functioning as it was supposed to. He had neck pain, headaches, decreased immune system, and Crohn's disease. His was one of the most devastating subluxation patterns there is, a reverse cervical curve.

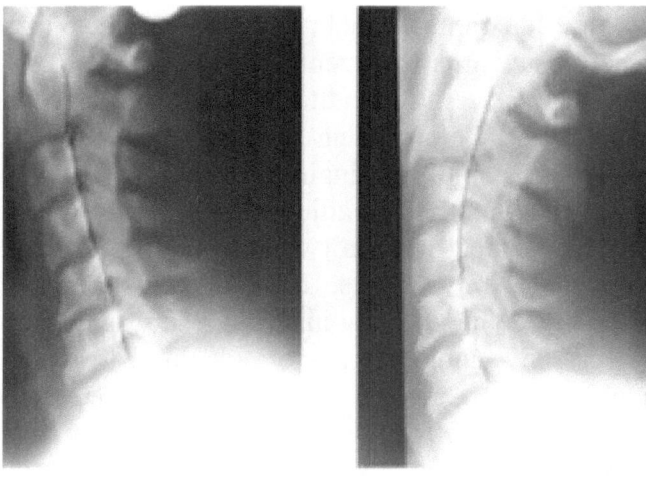

Before correction   After correction

John went through corrective care, and here are his results after one year. No more neck pain or headaches, and his immune system improved immensely.

The medical model says that adult scoliosis cannot be corrected. Well, the M.D.'s are partially right. Using drugs and surgery, scoliosis can't be corrected, but chiropractic corrective care can get phenomenal results. Here is a patient before and after two years of corrective care.

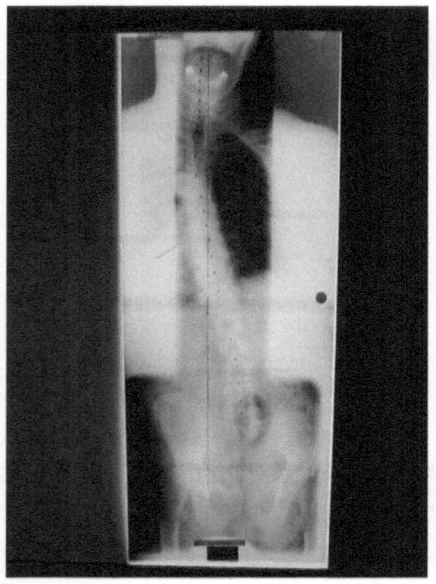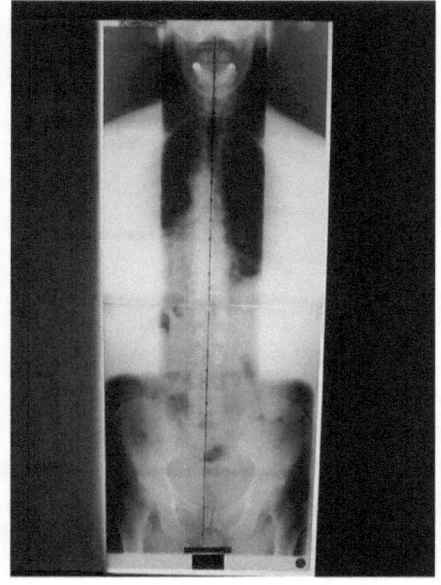

This patient came in with severe allergies and sinus problems.

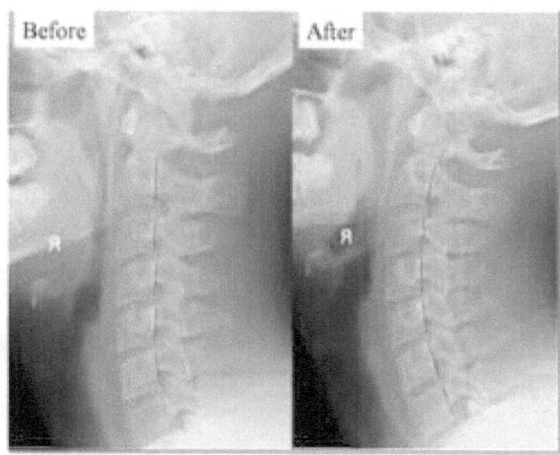

This is how he looked when I first saw him and how he looked three months later with all his symptoms resolved.

When you see a chiropractor, be patient and follow through with all the corrective care and wellness recommendations. The body heals in rhythm; each adjustment builds on the next. All these patients who had good results were disciplined and kept every scheduled appointment. Corrective chiropractic care may be a long process, but you can expect phenomenal results.

Step 3. Wellness Care

Once your body has fully healed and returned to homeostasis, routine wellness chiropractic care can keep your body in optimal condition and see that your health problems do not return. A wellness chiropractor can coach you on posture, nutrition, exercise, and stress management. Just like continuing an exercise program and eating well in order to sustain their benefits, we correct the spine to get back to wellness. When you make routine chiropractic care a part of your life, you avoid many of the aches and pains that so many people suffer, your joints last longer, and you can engage in more of the activities you love.

Myths and Facts about Chiropractic

As successful as chiropractic has become, there still are a lot of myths floating around. Times have definitely changed for the better, but many people still do not understand what chiropractors do. Let's talk about a few of the more common myths about chiropractic.

Chiropractors are not real doctors

Chiropractors are licensed as health care providers in every U.S. state and dozens of countries around the world. While the competition to attend chiropractic school is not as fierce as medical school, the chiropractic and medical school curricula are virtually identical. In fact, chiropractors have more hours of education than their medical counterparts. As part of their education, chiropractic students also complete approximately nine hundred hours of work in a clinical setting, assisting licensed chiropractors. Once chiropractic students graduate, they have to pass four sets of national board exams as well as state board exams in the states where they want to practice.

Chiropractors are not well educated.

Over my years in clinical practice a few patients have asked, "How long do you have to go to school to become a chiropractor?"

To receive a Doctor of Chiropractic, one must attend undergraduate school and receive a four-year degree and then apply to a chiropractic college. The Doctor of Chiropractic (D.C.) program is five years long and requires more than 4400 hours of classroom instruction. During the last 1½ years the student also works in the college chiropractic clinic delivering patient care and treatment. Five separate national board exams must be passed before licensing. The chart below compares a Doctor of Chiropractic (D.C.) degree program with a Medical Doctor (M.D.) degree program.

The following tables compare graduate school education hours for the D.C. and M.D. degrees.

Basic Science Education in Chiropractic and Medical Schools

Subject	Chiropractic Schools		Medical Schools	
	Hours	Percent of Total	Hours	Percent of Total
Anatomy	570	40	368	31
Biochemistry	150	11	120	10
Microbiology	120	8	120	10
Public Health	70	5	289	24
Physiology	305	21	142	12
Pathology	205	14	162	14
Total Hours	1,420	100	1,200	100

Curriculum Hours in Chiropractic and Medical Schools

Subject	Chiropractic Schools		Medical Schools	
	Mean	Percentage	Mean	Percentage
Basic science hours	1,416	29 percent	1,200	26 percent
Clinical science hours	3,406	71 percent	3467	74 percent
Total contact hours	4,822	100 percent	4,667	100 percent

Center for Studies in Health Policy, Washington, D.C. Personal communication from Meredith Gonyea, Ph.D., 1995.

The D.C. degree requires 4,822 hours of instruction; the M.D. degree, 4,667 hours.

It is clear that a Doctor of Chiropractic is highly qualified to care for all aspects of your health. We receive more education in musculoskeletal conditions than an M.D. Your first choice of a practitioner for many conditions should be a chiropractor.

Chiropractors receive extensive training, combined with many hours of practical work. Just like conventional medical doctors, chiropractors are medical professionals who are subject to testing, licensing, and monitoring by state and national peer-review boards. Federal and state programs such as Medicare, Medicaid, and Workers Compensation cover chiropractic, and all federal agencies accept sick-leave certificates signed by chiropractors.

The biggest difference between chiropractors and medical doctors lies not in their education or diagnostic ability but in their preferred method of treatment. Medical doctors are trained in the use of medicines (chemicals that affect your biochemistry) and surgery. Consequently, if you have an accident, trauma, or emergency, medical doctors can be very helpful. However, if your problem is that one of the bones in your spine is out of place, there is no chemical that can fix it. You need a physical treatment to correct a physical problem. That's where chiropractic really shines. Chiropractors use physical treatments—adjustments, exercises, stretches, traction—to treat conditions that are physical, rather than chemical, in origin, such as back pain, muscle spasms, headaches, and poor posture.

Medical doctors don't like chiropractors.

The American Medical Association's opposition to chiropractic was at its strongest in the 1940s under the leadership of Morris Fishbein. He called chiropractors "rabid dogs" and referred to them as "playful and cute, but killers." He tried to portray chiropractors as members of an unscientific cult, caring about nothing but taking their patients' money. Up to the late 1970s and early 1980s, the medical establishment purposely conspired to try to destroy the profession of chiropractic. In fact, a landmark lawsuit in the 1980s found that the American Medical Association was guilty of conspiracy, and the organization was ordered to pay restitution.

In the twenty years since, the position of most medical doctors has changed, primarily because of several studies that showed the superiority of chiropractic in treating a host of conditions, coupled with a better understanding among medical doctors about what chiropractors actually do. Many hospitals across the country now have chiropractors on staff, and many chiropractic offices have medical doctors on staff. Chiropractors and medical doctors are now more comfortable working together in cases where medical care is necessary as an adjunct to chiropractic care.

Once you start going to a chiropractor, you have to keep going the rest of your life.

This is a statement I frequently hear when the topic of chiropractic care comes up in conversation. It is partly true, maintaining your spine is obviously a good idea but you do not have to do this. You only have to continue going to the chiropractor as long as you wish to maintain the health of your neuromusculoskeletal system. Going to a chiropractor is much like going to the dentist, exercising at a gym, or eating a healthier diet: as long as you keep it up, you continue to enjoy the benefits.

Many years ago, dentists convinced everyone that the best time to go to the dentist is *before* your teeth hurt. Routine dental care will help your teeth remain healthy for a long time. Just like your teeth, your spine experiences normal wear and tear—you walk, drive, sit, lift, sleep, and bend. Regular chiropractic care can help you feel better, move more freely, and stay healthy throughout your lifetime. You can enjoy the benefits of chiropractic care even if you are only treated for a short time, but the real benefits come into play when you make chiropractic care a part of your lifestyle.

Chiropractic adjustments will cause you to have a stroke.

There are some medical doctors who still tell their patients to avoid chiropractic because sooner or later, they say, adjustments of the neck will cause a stroke. There is no valid scientific evidence for this claim. This is possible, especially if a patient is in a state of advanced arterial disease, but the risk is extremely small—about the same as being struck by lightning. In fact, you are 70,000 times more likely to suffer a stroke from the daily use of aspirin to prevent heart attacks than from a chiropractic adjustment. You are 37,000 times more likely to suffer a stroke for some unknown reason than from a chiropractic adjustment. When administered by a licensed Doctor of Chiropractic, adjustments are safe.

Frequently Asked Questions

What is a chiropractic adjustment?

The chiropractic adjustment is a gentle, quick thrust to a joint, typically in the spine, intended to restore normal position and movement. Adjustments are important for releasing adhesions in the joint and reducing stress on the nervous system. Because the nervous system is the master controller of the body, removing stress on the nervous system through chiropractic adjustments can lead to improved overall health.

How many adjustments will I need?

The length of treatment depends on five main factors.

- Your age
- Your general health
- The severity of your condition
- Your phase of spinal degeneration
- Your goal in treatment.

If you are young and in good health and have a mild condition that came on recently, you will need fewer adjustments than if you are older, in poor general health, and have been struggling with a problem for many years. The number of adjustments you will need also depends on whether you only want to alleviate your pain or whether you want to correct the problem and bring about optimal long-term health.

Will adjustments hurt?

Usually not. Some of my patients have experienced mild soreness after being adjusted, but this is more the exception than the rule. Most people feel better very quickly after an adjustment.

Have side effects or problems been reported from using chiropractic to treat back pain?

Patients may or may not experience side effects from chiropractic treatment. Effects may include localized discomfort, headache, or tiredness. These effects tend to be minor and to last only a day or two.

Do I still need to see the chiropractor after my pain is gone?

It is common for pain to disappear long before your condition is corrected. Pain is not a good indicator of health. People are often unaware of health problems because there is no pain associated with them. Heart disease and cancer, the two top killers, have no any symptoms until they are very advanced. The same is true of cavities in your teeth—there usually is no pain until a cavity becomes severe. Your pain may subside, but that doesn't mean your problem no longer exists. It is important to continue treatment so that the underlying cause of the pain can be corrected.

Routine chiropractic care is one of the simplest ways to maintain your health. Numerous research studies have shown that people who receive regular chiropractic care suffer fewer illnesses, injuries, and degenerative diseases, and they report a better quality of life.

The bottom line is that chiropractic care is a safe, effective treatment for a wide range of complaints, such as headaches, neck pain, low back pain, Carpal Tunnel Syndrome, Thoracic Outlet Syndrome, gastrointestinal complaints, wrist, elbow and shoulder pain, knee, hip and ankle pain, scoliosis, otitis media, and a host of other problems. While most of these disorders resolve within a few weeks or months, routine chiropractic care will help ensure optimal health and wellness for life.

Can chiropractors prescribe medication?

Chiropractors do not prescribe medications, but they refer patients to medical doctors when appropriate.

Chapter 3

Nutrition Basics

One on the most important decisions you make every single day is what food you put into your body. Food is not just an energy source—it contains hundreds or thousands of individual chemicals that influence a wide range of functions in your body, including your metabolic rate, immune function, emotional state, and, of course, body weight. Because of this, it is not good enough to simply count calories to lose weight. It is important to understand how food influences your body's biochemistry so that you can make informed choices. Once you understand some simple ideas about food, you can use this knowledge to help you lose weight and, more important, improve your overall health.

In this chapter, you will learn how your body uses the three basic types of food: carbohydrates, fats, and proteins. You will also learn the 10 Steps to Nutritional Health and Wellness™ and certain vitamins and minerals that are critical to successful health. If you put good things in, you will get good things out like health and vitality. If you choose to put bad things in, you will get bad things out: obesity, diabetes, acne, and ADHD, just to name a few. In this book I will teach you the secrets to health and wellness that you need to know to maintain optimal health.

Why Are We Fat?

There are a number of reasons why people are overweight and unhealthy. Some people think they struggle with nutrition because they don't know what is good for them, and many times they are just lying to themselves. I have given talks to small and large groups, and when I ask them what are healthy and unhealthy choices they always know the answer. But what proved this to me the most was when I did a health talk at a local elementary school here in Colorado. The first graders understood healthy food over unhealthy food even better than the adults. We all know that an apple is better than chocolate, but what we don't understand is *why*. Our body is made from millions of cells, and everything we do, from eating to moving and thinking, can change these cells. Health is cells functioning at 100 percent, and sickness is cells not functioning at 100 percent.

Everything you do affects your health; you are always moving toward health or away from it. Eating an apple moves you toward health, and I am sorry to tell you, ladies, that chocolate will move you away from health 100 percent of the time. You might not have symptoms from eating bad things now, but you will. The last time I checked, diabetes did not develop overnight. People generally have to work long and hard spending money and hours eating McDonald's to develop this! Is this really just a genetic problem?

Heart disease, cancers, cholesterol, and obesity are diseases of a poor lifestyle and are 100 percent preventable. People say they don't have the time to eat healthy food. Where did the time go? Did it expire? You always have time to eat healthily or at least make healthier choices: salad instead of French fries, water with lemon over soda, fruit over cake, small over large.

It's not hard to make better choices if you choose to eat out; but the best choice you can make is to make your food at home. Still others said they just can't seem to stay consistent in their diet. Let me tell you this. I can give you the exact foods you need to eat to be optimally healthy. I can also tell you the things to avoid. I can give you a hundred reasons to change your eating habits. I can and will give you the tools in this book to increase muscle, decrease body fat, improve your immune system, and live in a state of wellness. But what I can't give you is the true motivation or desire to change.

The fact that you're reading this book is a sign that you want to be on the road to wellness. A true desire to get and stay healthy is the first thing you have to acquire to make it happen. Health is a journey: it does not happen overnight. To be truly healthy you must eat healthy foods, exercise, and maintain a positive attitude. Being healthy does not mean that you lose weight by not eating or by taking some magic pill. It does not mean that you go on some weird fad diet and maybe lose fifty pounds. That's a number, not health; I bet you can't run five miles after you lose weight this way. Health is a choice, so I challenge you to make healthy choices and know that you are responsible for your health!

Diets Don't Work

If you type "diet" into a search engine, what do you find? You'll get more than 141,000,000 sites saying what new fad diet is right for you. Which one do you choose? How about none! To experience health, you must eat healthy, not diet. Most people know diets don't work or that they only last for a few days or weeks. That's why I don't diet. Instead, I have a scientifically proven, no-nonsense approach to nutrition.

America is one of the unhealthiest countries in the world with respect to our eating habits: we are a feed lot. There are a number of reasons we are so unhealthy, and the most-used excuse for this is convenience. It's easier to order a pizza or pull up to a drive-thru than it is to prepare a nutritious meal. This may be true, but the good news is, it's not as difficult as you may think to change from unhealthy to healthy. What you need to do is add!

That's right, I'm telling you that if you want to be healthy and lose weight, you need to add more food to what you're eating. How is that, you must be wondering. Well, it's quite simple: you need to add good choices to your meals before you take the bad  choices away. Remember what you just read about health and sickness—that you are moving toward health or away from it. By adding nutritious foods to your diet, you will move toward health 100 percent of the time. The best thing to start adding first is *fresh fiber!*

The key here is *fresh*, so add fresh fruits and veggies to every meal and eat them first! The darker the color, the better they are for you. If you add a salad before you eat a hamburger, you are moving toward health every time. The most important thing is adding fiber before you eat anything toxic. This will allow your body to absorb nutrients from the fiber and help your body digest the toxic substance and excrete it from your body.

It's that simple: you must add nutritious food before you take anything away. Once you have made fruits and vegetables a habit, your body will no longer crave bad food. This makes more sense than forbidding you to eat the foods you love that are bad for you. When you diet like that, you may last two weeks, but then you will pull into the first drive-thru you can find. Plus you will hate me when you can't eat the processed garbage that tastes so good. For a list of foods to add, see the second step in the 10 Steps for Nutritional Health and Wellness.

Metabolism Is Mission Control

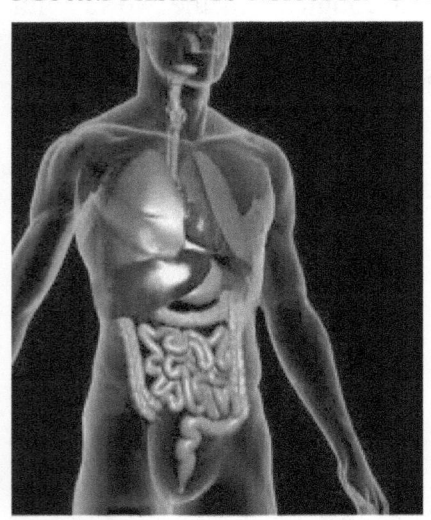

To improve how you feel and how you look on the outside, you have to improve what is going on inside. Depending on your metabolism, you are either a fat burner or a fat storer. The good news is that anyone can become a fat burner. You simply need to eat the right foods at the right times in the right combinations. This changes your metabolism, and metabolism is the secret to burning fat. Metabolism controls the breakdown of the nutrients you consume. There are two key factors to reprogramming your body to burn fat instead of storing it. First, you must increase your lean muscle mass (this is discussed in the next chapter). Second, if you want your body to go into fat-burning mode, you must eat smart.

Carbohydrates

Carbohydrate is the main fuel source that your body uses to think, run, walk, breathe, and do just about everything else. Next to water, it is the most consumed nutrient in the world. There are three types of carbohydrate: sugars, complex carbohydrates, and fiber.

In order for your body to use the sugars and starches in food, it must first break them down to a form that can be used by your cells. The first step in digestion occurs in your mouth, using an enzyme called *salivary amylase* that begins the process of breaking down starches into simple sugars. When food reaches the stomach, the digestion of carbohydrates stops. It begins again when your food leaves the stomach and enters the small intestine.

The main purpose of the digestion process is to convert the carbohydrate you consumed into a simple sugar called *glucose*, which is the primary fuel source for the brain, the central nervous system, and nearly every other cell in your body.

To ensure a readily available supply of glucose, the body stores it in the muscles and the liver in a form called *glycogen*. Glycogen is then converted back to glucose when your blood glucose level drops too low. If your body uses up all its glycogen, it will start breaking down muscle in order to provide your vital organs with the glucose they need to function.

The two hormones that regulate the level of glucose in your blood are *insulin* and *glucagon*. Insulin is a hormone that is released when your blood glucose levels rise, as typically occurs after you eat food containing carbohydrates. Its function is to signal the liver and muscle cells to remove excess glucose from the blood and store it as glycogen.

Glucagon has the opposite effect. When your blood glucose levels are too low, glucagon signals the muscles and liver to convert glycogen back to glucose and release it into the bloodstream. The balance of these two hormones keeps blood glucose levels within a fairly narrow range.

There are some instances where the body is unable to maintain healthy blood glucose levels. The most common condition, diabetes, is caused by a loss of normal insulin function. Those with diabetes have an abnormally high blood glucose level. A much rarer condition called hypoglycemia occurs when blood glucose levels are abnormally low.

Not all carbohydrates have the same effect on blood glucose levels. Starches are much larger molecules than sugars and therefore take longer to break down and enter the bloodstream as glucose. Sugars, on the other hand, are simple molecules that are quickly converted into glucose. Sugars tend to create a sharp spike in blood glucose levels, whereas starches cause a more gradual increase.

The ability of a food to elevate blood glucose levels is referred to as its *glycemic index*. Simple sugars have a high glycemic index because they cause a rapid increase in blood glucose levels, but starches have a low glycemic index because the increase is slower.

Cells in the body can shut down and no longer allow insulin into the cells. This *insulin resistance* is a major problem for many Americans, because the typical American diet is full of processed foods and cereal grains, all high glycemic foods.

Insulin resistance also leads to pancreatic failure as the pancreas works too hard trying to produce more and more insulin to get its job done. This affects homeostasis of the immune system as well, because the molecules of glucose and Vitamin C are almost identical and the body cannot tell them apart. This confusion leads to lowered immunity and more sickness. Insulin resistance also affects the ability of the bones to absorb calcium. People who are insulin resistant can supplement with any amount of calcium and it will just be excreted in their urine. Americans consume the most milk in the world, but they have the highest rate of osteoporosis.

High glycemic index foods will tend to increase the storage of body fat because each fat cell can respond to insulin, remove glucose from the blood, and store it, but instead of storing the extra glucose as glycogen like the muscles and liver, it stores the excess glucose as fat. The higher your blood glucose, the more will be stored in your fat cells. To keep from storing extra glucose and enlarging your fat cells, it is important to consume low glycemic index foods, especially fruits and vegetables.

Here are some examples of foods having a high and low glycemic index.

High Glycemic Index FoodsLow Glycemic Index Foods
Fried foodsFruits
White breadVegetables
White RiceNuts
White Sugar
Snack Foods
Pop (soda's diet or regular)
Cereal grains

Proteins

Proteins are required to maintain the normal structure and function of the body. Whereas carbohydrate in the form of glucose is the primary fuel source for the body, proteins are its primary building blocks. Muscle, bone, and connective tissue are made from protein, as are enzymes, antibodies, hemoglobin and even DNA.

Proteins are made up of twenty different amino acids. Twelve of them, the *nonessential amino acids*, can be synthesized in your body and do not need to come from your diet. The other eight, the *essential amino acids,* must come from your diet, because your body cannot make them.

The essential amino acids are *isoleucine, leucine, lysine, methionine, phenylalanine, threonine, tryptophan,* and *valine.* If you don't get enough of these amino acids in your diet, your body can't repair itself, your immune system can't do its job properly, your metabolism slows down and you tend to feel sluggish, depressed, and tired. The primary sources of these amino acids are protein foods such as meat, fish, eggs, beans, and legumes.

Before your body can use the protein in your food, it must first break it down to its individual amino acids. Digestion of protein begins in the stomach, where acids and *proteolytic enzymes* begin the process of releasing amino acids. Some amino acids are absorbed directly through the stomach lining and enter the bloodstream. The remaining protein then enters the small intestine, where digestion is completed.

Once the amino acids enter the blood, the body can use them to build red blood cells, muscle tissue, immune factors, and whatever else is needed, but there's a catch. All twenty amino acids must be present in the blood in the proper ratios if they are to manufacture (or synthesize) new proteins. If one amino acid is missing or inadequate, it becomes the limiting factor for protein synthesis. Eating higher biological value proteins will give you the greatest amount of usable protein for the number of calories consumed.

The proteins in some foods, such as eggs and meat, have amino acid profiles that closely match the ratios in the human body and are said to have a high *biological value (BV).* Eggs have the highest BV and are used as the standard by which all other proteins are measured. In the Biological Value of Proteins chart, you will see that eggs have a BV score of 100, meaning that approximately 100 percent of the protein in eggs can be metabolized by the body.

Lactose-free whey protein concentrate has a BV of 104. Can more than 100 percent of its proteins be absorbed? Well, no. When the standard of biological value was initially set, egg protein had the highest known profile, so eggs were made the standard at 100. Since then it has been found that Lactose-free whey protein has a slightly better profile than eggs, and instead of changing the standard to whey, it was decided o give it a higher score than eggs. Other protein foods with high BVs are whole beans 96, human milk 95 (not cows' milk), brown long grain rice 83, and beef 80. Other foods, such as grains and vegetables, have amino acid profiles that do not match the body's needs very well. These are lower-quality proteins and have lower BV scores.

The other measure of the quality of protein is how much of it your body can digest—its *Percentage of Digestion (PD)* score. When you cut back on your food intake in an effort to lose weight, it is important to eat higher PD proteins, as well as those with the highest BV score.

Proteins have an added advantage in that they don't cause a rapid increase in blood glucose levels, making them low glycemic index foods. In addition, proteins increase the body's metabolism more than carbohydrates and fats, and they provide the building blocks for mood-elevating neurotransmitters such as *phenylalanine* and *tryptophan*.

Proteins are critical for building lean muscle tissue, and for maintaining stable blood glucose levels, immune function, and normal brain chemistry. When trying to lose weight, it is important to increase your intake of protein to help protect your muscle and bone tissue and to boost your metabolism during periods of calorie restriction.

Fats

Of the three major components of food, fats are the most misunderstood and the most vilified. Fat is not the bad thing it's made out to be. In fact, all the cells in your body are surrounded by a membrane of fat. That's why you can go swimming and not dissolve in the water. Your brain, spinal cord, and nervous system are largely made from fat, as are many of your hormones, such as testosterone and estrogen. Fat provides your body with energy and insulation from the cold, and it protects your organs from damage. Next to water, fat is the most abundant substance in the human body, ideally averaging about 10 to 20 percent of a person's total body weight.

Dietary fats are necessary for the proper absorption of fat-soluble vitamins, and scientists recently discovered that some fats in the diet are used for sending signals to the brain to control how much you eat. It is not fats themselves that you should avoid. You should only avoid eating too much of the wrong kind: saturated fat.

Dietary fats come in four main forms: saturated fats, polyunsaturated fats, monounsaturated fats, and cholesterol.

Saturated fats are the demons of dietary fats. They can elevate blood cholesterol and contribute to the development of heart disease, and they also tend to cause a low-grade inflammation in the body. Snack foods contain the most common saturated fats, such as palm oil and palm kernel oil. Common foods that are high in saturated fat and that should be avoided are:

- Cream, half-and-half, and whole-milk dairy products such as cheese, ice cream, and sour cream

- Processed grain products, such as cookies, cakes, muffins, pastries, cereals, and fried foods

It is a misconception that all saturated fats are bad for you, however. There are some foods that are high in saturated fat and nevertheless do wonders for your body and your health, such as coconut and coconut oil, egg yolk, and avocadoes.

Polyunsaturated fats are found in seafood and certain oils. They remain liquid at room temperature, whereas saturated fats are usually solid. Unlike saturated fats, polyunsaturates, especially Omega 3 fatty acids found in fish oil, will help to decrease serum cholesterol, both LDL and HDL, and will help to decrease inflammation in the body.

Monounsaturated fats are even healthier than polyunsaturated fats. They not only help decrease your LDL cholesterol, but they also help raise your HDL cholesterol. Grape seed oil, flaxseed oil, and olive oil are good sources of monounsaturated fats, as are avocados and nuts. The best oils for cooking are coconut oil and almond oil, as they both can tolerate high heat and will not break down at high heat as olive oil does.

Cholesterol is a waxy fat that is found exclusively in animal foods: beef, chicken, fish, turkey, eggs, dairy, and others. Years ago it was believed that consuming cholesterol in your diet led to an increase in your blood cholesterol, but this turned out not to be the case.

In fact, LDL cholesterol is vital to your body when an injury occurs. LDL is the main building block used to repair internal organs when they are damaged and our bodies are under stress.

High cholesterol is accompanied by an excess of *cortisol* in the body. Cortisol is a steroid hormone released by the adrenal glands in response to stress. That is why nearly half of America's population has high cholesterol levels—cortisol causes less LDL to be taken out of the bloodstream, raising our levels of cholesterol. The major lifestyle contributors to high cholesterol are saturated fat, overweight, and lack of exercise.

Calories

A calorie is a measure of the energy content of food. Energy is what your body uses to keep your heart pumping, keep your lungs expanding, and allow your mind to think and your muscles to move. Carbohydrates and proteins provide four calories per gram, fats provide nine calories per gram, and alcohol provides seven calories per gram. Ounce for ounce, proteins and carbohydrates give you half as many calories as fat. It is important to remember that calories are not your enemy. As strange as it may seem, if you eat too little food, it is harder to lose weight, because too much calorie restriction slows down your metabolism and causes your body to store fat.

Maintaining a healthy body composition is a balancing act between the calories you consume and the calories you burn. We have discussed the sources of energy in food: carbohydrates, proteins, and fats. Let's look at the other side of the equation: how your body burns the calories you consume.

You burn calories in two ways: Through metabolism, measured as *basal energy expenditure (BEE)* and by being active. The biggest user of calories is metabolism, which provides the energy your organs use to keep your body alive and to build muscle, bone, and connective tissue. Metabolism burns up to 75 percent of the day's calories. Activity burns anywhere from 15 percent to 35 percent.

The most effective way to achieve healthy fat loss is to combine a moderate decrease in calorie intake with an increase in energy expenditure.

Get Your FACTS Straight: The 10 Steps for Nutritional Health and Wellness

If you eat the right *Foods* in the right *Amounts* in the right *Combinations* at the right *Times,* you can't help but become healthier. We have helped hundreds of people lose up to 50 pounds and more while they improve their health and self-esteem by following our 10 Steps for Nutritional Health and Wellness.

Step 1. Add Before You Take Away

You need to add fresh fiber to every meal and eat it first. The first thing in is the most important, and there is nothing better than fiber for digestion and absorption. This will make a big difference to your health, guaranteed. After you've established the habit of eating fiber, you will no longer crave fatty, sugary, processed foods. If you know that at lunch you will eat fast food, make sure to bring fruit or veggies with you and eat it beforehand. If you are under stress and crave chocolate, try this: before you eat that chocolate, eat three whole carrots.

Step 2. Eat Six Times a Day—Eat More To Weigh Less

Did you know Sumo wrestlers in training eat only once a day? That's right. In the middle of the day, Sumo wrestlers consume a large buffet-style meal. They do this because eating once a day and overeating slows metabolism, causing you to gain weight. This may be advantageous for Sumo wrestling, so many Americans are training to become Sumo wrestlers and don't even know it. They have coffee in the morning, maybe skip lunch because they're busy, and go home and eat a huge dinner and fall asleep on the couch. The number of times you eat in a day is critical to your health and wellness. What kind of nutritional decision do you make when you miss lunch and go home starving? Does drive- thru, pizza, or fast food sound like you?

Eating right means more than eating three square meals a day. In fact, if you eat three meals a day you probably have a slow, inefficient metabolism. Clinical research now shows that to have an optimally functioning metabolism that can increase energy and burn fat, you should be eating *every three hours*. A study published in the *European Journal of Clinical Nutrition* found that the average resting metabolic rate of people who ate six times a day was much faster than that of people who consumed just three meals a day.

Eating six meals a day also allows your body to maintain lean muscle mass. The more lean muscle you have, the faster your metabolism. A study in *The New England Journal of Medicine* showed that, in only two weeks, people who ate the same amount of food in frequent portion-controlled meals as opposed to three larger meals reduced their LDL cholesterol levels almost 15 percent, reduced their blood cortisol levels by more than 17 percent, and reduced their insulin levels by almost 28 percent.

A study called *Effects of Meal Frequency on Body Composition* showed that people who ate two meals a day lost twice as much lean muscle and half as much fat as people who ate six meals a day.

Step 3. The Power of Protein and Fat

This nutrition plan is not a high-protein, low-carbohydrate diet. Such a diet can create a process in the body called *ketosis*, which the body breaks down into ketones. This kind of diet can cause a number of health problems, including reduced kidney function, constipation, and chronic fatigue.

The government's daily value (DV) for protein is almost 50 grams a day for women and 60 for men. This isn't very much. The amount of protein you need to eat each day should minimally be half your body weight in grams. If you are 200 lbs, then you should be eating approximately 100 grams of protein. If you are exercising you should eat your body weight in protein. If you weigh 150 lbs, you should be eating 150 grams of protein.

Most Americans are deficient in the amount of protein needed for the body's metabolism to be efficient. A study at Texas Christian Women's University by Scott Connelly, M.D., put two groups of middle-aged obese women on a low-calorie diet and strength training program. Both groups consumed the same number of calories, but one group was on a high carbohydrate-low protein diet, while the other was a high protein-low carbohydrate diet. Both groups lost weight, but the low protein group lost 65 percent fat, 35 percent lean mass. The women on the high protein diet lost almost exclusively fat.

Fat is not your enemy when you are eating healthy fats. We need fat as we need protein and carbohydrates, but it needs to be the right fat. Consider this: What do you feed a cow to make it fat? Grains.

Weeks or months prior to slaughter, meat cows are given grains to put on weight. We are just like cows when we consume processed grains high in sugar. Again, fat is not the enemy; grains are. Another enemy brings us back to the topic of cows. We are the only species to drink milk past weaning and the only species to drink another animal's milk. When was the last time you saw a tiger drinking a hippo's milk? Before the introduction to milk, we never had acne, osteoporosis, cancers, obesity, and diabetes.

When I'm asked how to get calcium without drinking milk, I ask in return, "What do elephants eat for calcium?" They eat greens, and so can you. The darker the veggie, the better it is for you. The best one is broccoli—eat some every day. Dairy foods are acidic, and they take calcium out of our bones. I have yet to read a peer-reviewed article that has any evidence to support that dairy products supply calcium to our body. I think everyone needs to read an article by Dr. Robert Kradjian called "The Milk Letter." It can be found at http://www.notmilk.com/kradjian.html.

Step 4. Eat the Right Combinations to Stabilize Blood Sugar

Your brain needs carbohydrates in order to have the glucose it needs to function. A 2003 study published in the *Journal of Nutrition* found that you need not give up carbohydrates to stabilize blood sugar. A smarter plan, according to these researchers, is to balance carbohydrates with protein in every meal. A study in *The Journal of Physiology and Behavior* investigated the effects of different carbohydrate-to-protein ratios on cognitive performance; the subjects' diets were carbohydrate-rich, balanced, and protein-rich. The study found that balanced eating resulted in better brain functioning. Eating frequently and in the right combinations has been proven to stabilize blood sugar, which will stabilize your energy levels throughout the day. Each meal should contain the following:

Protein (vegetables and lean, grass-fed) 20 to 30 percent

Carbohydrates (fruits and mostly vegetables) 40 to 50 percent

Fat (meats and fish; vegetables; nuts) 30 to 40 percent

Health and Wellness Guide to Healthy Living

Food Guide

Proteins: Vegetables/Organic
Beef/Fish 20-30 percent
Chicken/Turkey/Eggs

Fat: Organic Meats/Fish/Coconut
Oil 30-40 percentNuts/Avocado

Carbohydrates: Fruits/Vegetables
40-50 percent

Step 5: Hydrate

Water is the single most abundant nutrient in the body, accounting for 60 to 65 percent of your total weight, and it is the least forgiving of all the nutrients you consume. You can survive for weeks without food, but for only a couple of days without water. Water is responsible for the transport of nutrients, oxygen, and waste products, and it regulates your body temperature and serves as the medium in which your body's chemical reactions take place.

Most people do not drink enough water. When you begin an exercise and diet program, it is very important to consume enough water. How much water should you drink? Each day, you will want to consume at least 1.5 to 2 liters of water or half your body weight in ounces. A 200-lb. person should be consuming 100 ounces of water per day. This sounds like a lot, but it really isn't. The trick is to get a liter-sized bottle and carry it with you. Make it a goal to drink two bottles of water every day. Coffee, tea, milk, juice, sports drinks, and the like do not count toward the total amount of water. Ice water has the added benefit of having to be warmed up when it is consumed, causing your body to expend more calories heating it. Who says you can't get something for nothing?

If you are not getting enough water, your body will react by pulling it from other places, including your blood. This causes some smaller vessels, the capillaries, to close, making your blood thicker, more susceptible to clotting, and harder to pump through your system. This can have serious implications in hypertension, high cholesterol, and heart disease. Recent studies have also linked the lack of water to headaches, arthritis, and heartburn.

Have you have ever gotten up in the morning feeling bloated, or tried on a ring or a shoe that fit yesterday but is too tight to wear today? Chances are your body is trying to tell you something. If you have a problem with water retention, excess salt may be the cause.

Your body will tolerate a certain amount of sodium, but the more salt you consume, the more fluid you need to drink to dilute it.

What if I told you that being dehydrated promotes the increase of body fat? Water, like glycogen, contributes to energy storage. Without water, extra glucose remains in the bloodstream until it reaches the liver and it is stored as fat. Your body takes water from inside cells, including fat cells, in an effort to compensate for a dehydrated state. Less water in your fat cells means less mobilization of fat for energy.

One of the liver's primary functions is to metabolize stored fat into energy. The kidneys are responsible for filtering toxins, wastes, ingested water, and salts out of the bloodstream. If you are dehydrated, the kidneys cannot function properly, and the liver must work overtime to compensate. As a result, it metabolizes less fat. Reducing water in the body as little as 5 percent can result in as much as a 20 to 30 percent drop in your physical performance. A 10 percent reduction can make you sick, and 20 percent can cause death.

Step 6: Take a Nutritional Supplement

Some say we don't need to take supplements because the human body doesn't need them. As long as you eat a balanced diet, you can get everything your body needs. While it is certainly true that people living a thousand years ago did not have multivitamins, they also did not have toxic chemicals pumped into their environment every year; they were not exposed to a pervasive man-made electromagnetic field from power lines and cell phones; they did not eat processed foods that contained artificial colors, flavors, and preservatives; they were not sedentary; and they were not under constant stress. Our bodies were not designed for a fast-paced, high-stress, processed lifestyle.

The reality is that we need to give our body some help in order to stay healthy in the world today. That's where supplements come in.

Vitamins and Minerals

A colleague of mine told me about a challenge that one of his professors made to the class during a graduate nutrition course he took at the University of Minnesota. The challenge was to construct a 2000 calorie-per-day diet that at least met the Recommended Dietary Allowances (RDA) for vitamins and minerals without the use of supplements. After all, we have always heard that if you eat a well-balanced diet, you don't need to take vitamin supplements, right? The professor was putting that statement to the test.

To everyone's surprise, no one was able to come up with a sustainable daily diet that met the minimum requirements for mineral intake. The problem was not with getting the minimum vitamin intake; that was relatively easy. The challenge was getting enough of a few very important minerals, especially zinc. Unless you eat oysters or dark turkey meat every day, it is impossible to get the minimum daily value of zinc through diet alone.

So, it is not possible to get everything we need from the food we eat. But how could this be? Throughout history humans must have been able to get everything they needed from their diet in order to maintain the population. The answer has to do with modern farming techniques, fertilizers, and environmental stresses.

Following the Second World War, chemical manufacturers were sitting on huge stockpiles of phosphates and nitrates that were intended for use in explosives. They discovered that when they spread these same phosphates and nitrates on the soil where plants were growing, the plants grew bigger and looked healthier. Thus began the fertilizer industry.

The problem with modern fertilizers is that they don't replace soil trace minerals, such as chromium, zinc, and copper, as do cow manure and other natural fertilizers. Over time, these trace minerals become more and more depleted from the soil, and consequently our food supply becomes more depleted as well. The bottom line is that in order to get enough trace minerals in our diet to at least meet the minimum RDAs, it is necessary to take a good quality supplement.

There is substantial evidence that taking doses of a class of nutrients called antioxidants (especially vitamins A, C, E, and selenium) that far exceed the RDA minimums can help prevent heart disease, mitigate some of the detrimental effects of environmental pollutants, and promote healthy immune function.

How to Select a Good Multivitamin

All vitamin supplements are not created equal. Supplements are just like anything else—there are some good ones and a whole lot that are not as good. Here are a few keys to determining whether a particular vitamin is good.

- In general, supplements sold through health care professionals are top quality. They tend to be a little more expensive than the supplements you find at your local drug store because the ingredients that go into them tend to be of a higher quality.

- Good quality vitamins have chelated minerals. This makes a huge difference in how well the minerals are absorbed by your body. For example, in supplements whose calcium source is calcium carbonate, less than 25 percent of the calcium is absorbed. In older adults, this absorption rate drops to about 10 percent. In contrast, if you take a supplement with calcium citrate (chelated), 30 to 50 percent of it is absorbed—almost twice as much! If you have any questions about specific supplements, be sure to ask your wellness chiropractor or nutritionist.

- Good quality vitamins require that you take more than one capsule or tablet per day. This is simply because good quality ingredients take up more space than their cheaper counterparts. Depending on your individual needs, you could be taking anywhere from two tablets per day for general nutritional support to six tablets per day if you are an athlete or have special nutritional needs.

- Good quality vitamins will have adequate levels of biotin, whereas cheap drugstore vitamins will not. Biotin is the most expensive vitamin to manufacture. Cheaper vitamins will have very little biotin in them compared to higher quality vitamins, and is one of the reasons why good vitamins cost a little more.

Step 7. Check Your Blood pH

In my opinion, Dr. Morter, M.D. helped discovered one of the greatest breakthroughs in health assessment in the last hundred years and that is that all organic matter has a potential of hydrogen (pH) level, including humans. Our body's pH is an indicator of our health status. There are a number of ways to test pH: some test saliva, others test urine, and others use more invasive measures. But the one thing all experts agree on is that the blood pH needs to remain between 7.35 and 7.45 to avoid serious health problems. Just as the body temperature is kept at precisely 98.6 degrees, the blood and tissues must be kept in a very narrow pH range.

Many factors influence pH, including stress, certain medications, subluxation, and diet. Acidity is a major stressor to the body; it can send the body into survival mode to maintain homeostasis. Imbalance in blood pH sets the stage for disease. An unbalanced diet high in acid-producing foods such as animal protein, sugar, caffeine, and processed foods puts pressure on the body's ability to maintain pH neutrality. The extra effort required can deplete the body of alkaline minerals such as sodium, potassium, magnesium, and calcium, making a person prone to chronic and degenerative disease. Minerals are borrowed from vital organs and bones to neutralize the acid and safely remove it from the body. Because of this, the body can suffer severe and prolonged damage that may go undetected for years.

Pathogenic microforms thrive in acidity, and most diseases follow microform overgrowth. General signs of microform overgrowth include pain, infection, fatigue, indigestion, diarrhea, intestinal trouble, depression, hyperactivity, asthma, hemorrhoids, colds and flu, respiratory problems, dry itchy skin, receding gums, dizziness, joint pain, bad breath, ulcers, colitis, heartburn, dry mouth, menstrual problems, irritability, puffy eyes, lack of sex drive, skin rash, hives, hormonal imbalance, vaginal yeast infection, cysts and tumors, bladder infections and allergies. If our pH gets too far out of balance, our bodies will leach the calcium, which is alkaline, out of our bones and teeth to correct an acidic condition.

It is important to consume at least 60 percent alkaline-producing foods in order to maintain health. We need plenty of (alkaline-producing) fresh fruits and vegetables to balance our necessary (acid-producing) protein intake. This is another reason to avoid processed, sugary, and simple-carbohydrate foods, which are all acid-producing.

FOOD CATEGORY	High Alkaline	Alkaline	Low Alkaline	Low Acid	Acid	High Acid
BEANS, VEGETABLES, LEGUMES	Parsley, Raw Spinach, Broccoli, Celery, Garlic, Barley Grass	Carrots, Green Beans, Lima Beans, Beets, Lettuce, Zucchini, Carob	Squash, Asparagus, Tomato, Rhubarb, Fresh Corn, Mushrooms, Onions, Cabbage, Peas, Cauliflower,	Sweet Potato, Cooked Spinach, Kidney Beans	Pinto Beans, Navy Beans	Pickled Vegetables

FOOD CATEGORY	High Alkaline	Alkaline	Low Alkaline	Low Acid	Acid	High Acid
			Turnip, Beetroot, Potato, Olives, Soybeans, Tofu			
FRUIT	Dried Figs, Raisins	Dates, Blackcurrant, Grapes, Papaya, Kiwi, Berries, Apples, Pears	Coconut, Sour Cherries, Oranges, Cherries, Pineapple, Peaches, Avocados, Grapefruit, Mangoes, Strawberries, Papayas, Lemons, Watermelon, Limes	Blueberries, Cranberries, Bananas, Plums, Processed Fruit Juices	Canned Fruit	
GRAINS, CEREALS			Amaranth, Lentils, Sweet Corn, Wild Rice, Quinoa, Millet, Buckwheat	Rye Bread, Whole Grain Bread, Oats, Brown Rice	White Rice, White Bread, Pastries, Biscuits, Pasta	
MEAT				Liver, Oysters, Organ Meat	Fish, Turkey, Chicken, Lamb	Beef, Pork, Veal, Shellfish, Canned Tuna & Sardines
EGGS & DAIRY		Breast Milk	Soy Cheese, Soy Milk, Goat Milk, Goat Cheese, Buttermilk, Whey	Whole Milk, Butter, Yogurt, Cottage Cheese, Cream, Ice Cream	Eggs, Camembert, Hard Cheese	Parmesan, Processed Cheese
NUTS & SEEDS		Hazelnuts, Almonds	Chestnuts, Brazils, Coconut	Pumpkin, Sesame, Sunflower Seeds	Pecans, Cashews, Pistachios	Peanuts, Walnuts

FOOD CATEGORY	High Alkaline	Alkaline	Low Alkaline	Low Acid	Acid	High Acid
OILS		Coconut Oil	Flax Seed Oil, Olive Oil, Canola Oil	Corn Oil, Sunflower Oil, Margarine, Lard		
BEVERAGES	Herb Teas, Lemon Water	Green Tea	Ginger Tea	Cocoa	Wine, Soda/Pop	Tea (black), Coffee, Beer, Liquor
SWEETENERS, CONDIMENTS	Stevia	Maple Syrup, Rice Syrup	Raw Honey, Raw Sugar	White Sugar, Processed Honey	Milk Chocolate, Brown Sugar, Molasses, Jam, Ketchup, Mayonnaise, Mustard, Vinegar	Artificial Sweeteners

Getting Your Greens

We know that we should eat more fruits and vegetables, but we prefer easily available fast foods and snack foods. As a result, we have lost our taste for fruits and vegetables. It is not uncommon for people to go for weeks without consuming a single serving of fresh vegetables. This is not good. The human body is designed to live on a diet high in fruits and vegetables, and it is dependent on compounds unique to plant foods in order to operate correctly.

Just as vitamins and minerals are critical for good health, so are a group of compounds from fruits and vegetables called *phytonutrients*, the brightly colored pigments that are the engines of life for plants. The best-known phytonutrients are the *carotenoids* and *flavonoids*. As a group, they are called *polyphenols*.

Carotenoids, the bright yellow, orange, and red pigments, are found in vividly colored vegetables such as carrots, beets, and sweet potatoes.

Carotenoid	Common Food Source
alpha-carotene	carrots
beta-carotene	dark green and yellow vegetables (e.g., broccoli, sweet potatoes, pumpkin, carrots)
beta-cryptoxanthin	citrus, peaches, apricots
lutein	leafy greens (e.g., kale, spinach, turnip greens)
lycopene	tomato products, pink grapefruit, watermelon, guava
zeaxanthin	green vegetables, eggs, citrus

Flavonoids are reddish pigments found in red grape skins and citrus fruits. Food sources rich in polyphenols include onions, apples, tea, red wine, red grapes, grape juice, strawberries, raspberries, blueberries, cranberries, and certain nuts.

Nonflavonoids	Sources
ellagic acid	strawberries, blueberries, raspberries
coumarins	

Flavonoids	Sources
anthocyanins	fruits
catechins	tea, wine
flavanones	citrus
flavones	fruits and vegetables
flavonols	fruits, vegetables, tea, wine
isoflavones	soybeans

Phytonutrients are powerful antioxidants that help reduce your risk of cancer and heart disease, and they are important for healthy immune function. This protective mechanism explains why cultures whose diets are rich in fruits and vegetables, such as the Mediterranean diet, have the lowest rates of cancer, heart disease, and degenerative disease. The importance of fruits and vegetables is no secret, but most of us still don't eat enough of them.

The Top Phytonutrient Foods

While nearly all plant foods contain health-promoting phytochemicals, the following are the most phyto-dense food sources.

- tomatoes
- broccoli
- flax seeds
- citrus fruits
- blueberries
- sweet potatoes, yams
- chili peppers
- legumes: beans and lentils
- green tea
- red grapes
- carrots
- kale
- squash
- spinach

Real fruits and vegetables are better, but if you find it difficult to get enough greens in your diet, you should take a greens supplement. These are usually in the form of a powder mixed with water or juice and consumed at least once a day. Taking a greens supplement is much like taking a multivitamin—it is a simple way to ensure that your body has everything it needs in order to be healthy.

The greens blend I use, called Chiropractors Blend Natural Greens, also contains wheat grass. This supplement improves the pH of the blood and has the antioxidant power of ten servings of fruits and vegetables in every scoop. For those who are allergic to wheat, there is a product called Greens First: Doctors for Nutrition. For more information on pH, see *The pH Miracle* by Robert Young, PhD.

Step 8. Omega 3 Fatty Acids

Omega 3 fatty acids are polyunsaturated fats found in the oils of some fish and in plants such as flax, walnuts, and hemp. Nutritionally, the four most important fatty acids are *alpha linolenic acid (ALA), linolenic acid (LNA), eicosapentaenoic acid (EPA)*, and *docosahexaenoic acid (DHA)*. These are classified as *essential* fatty acids because the body cannot synthesize them—they must be obtained from your diet. They are crucial to the development of the human brain.

LNA and ANA come from vegetarian sources such as flax, and EPA and DHA come from fish and wild game. The typical American diet is almost devoid of the Omega 3s. In fact, researchers believe that about 60 percent of Americans are deficient in Omega 3 fatty acids, and about 20 percent have so little that test methods cannot detect any in their blood. We are also toxic in Omega 6, and this in combination with Omega 3 deficiencies leads to a greater chance of illness in people of all ages. You need to maintain a ratio of 1:1 between Omega 3 and Omega 6, but the typical American ratio is 20:1 Omega 6 to Omega 3. It is clear that supplementation with Omega 3 is essential.

Omega 3 deficiencies have been tied to many conditions, including skin conditions, dyslexia, hyperactivity, learning disorders, depression, memory problems, allergies, heart disease, and chronic inflammation

The most natural way to increase your intake of Omega 3 fatty acids is to consume more fish and wild game. Unfortunately, fish now contains toxic heavy metals and industrial chemicals, so obtaining Omega 3 fatty acids from food is no longer the best option. The best source of Omega 3 is bottled fish oil. When selecting good fish oils, though, make sure that it goes through a process called "molecular distillation" and is bottled at the source.

The safest and most convenient way to add more Omega 3s to your diet is to take a high-quality daily supplement. The purest form of fish oil I have found comes from Norway through Canada and can be purchased from *www.innatechoice.com* or on my Web site at *www.drtomalin.com.* Innate Choice Omega Sufficiency fish oil is the world's premier EPA/DHA Omega 3 fish oil supplement. It is 100 percent natural, triple molecularly distilled, ultra purified wild anchovy, herring, and sardine oil that is infused with antioxidants before bottling in Norway.

Step 9: Add Probiotics

Inside each of us live vast numbers of beneficial bacteria that are vital to good health. Our gastrointestinal tract is home to more than 400 different species of bacteria that perform important functions in the body. Due to emotional stress, prescription drug use, environmental toxins, and poor diet, these helpful bacteria are destroyed and must be replenished through the use of *probiotics* if we are to be healthy.

Probiotic, meaning *for life*, is now mostly used to refer to concentrated supplements of beneficial bacteria needed by humans and animals. Friendly bacteria promote the body's natural immunity, manufacture B vitamins such as biotin, niacin (B3), pyridoxine (B6) and folic acid; act as anticarcinogenic (anticancer) factors; act as watchdogs against the spread of many harmful bacteria, viruses and fungi; help to reduce cholesterol levels; help protect against the negative effects of radiation and toxic pollutants, enhancing immune function; enhance bowel function; and generally help to keep us healthy.

How to Select a Good Probiotic

It is important to choose carefully when selecting a probiotic supplement. Spending extra time looking for the right product and spending extra money purchasing the right product will pay dividends of better health in the long run. Here are a few things to look for when selecting a probiotic supplement.

- *Number of Organisms.* Product should say guarantee the number of viable organisms: at least 1 billion organisms per gram is required for a therapeutic dosage.

- *Type of Organisms.* The most therapeutically important types of bacteria are *Lactobacillus acidophilus, Lactobacillus bulgaricus, Streptococcus thermophilus* and *Bifidobacterium bifidum.* L. bulgaricus and S. thermophilus are useful for encouraging the growth of B. bifidum in the intestines. INT 9, DDS-1, and NAS strains of L. acidophilus are all good strains to use. For more information on probiotics, ask your wellness chiropractor or nutritionist.

- *Absence of Dairy Ingredients.* You need to make sure the capsule is dairy-free in order to get full absorption. When the supplement is taken in yogurt or with milk, absorption is blocked by the mucus they form.

Step 10. Get Adjusted!

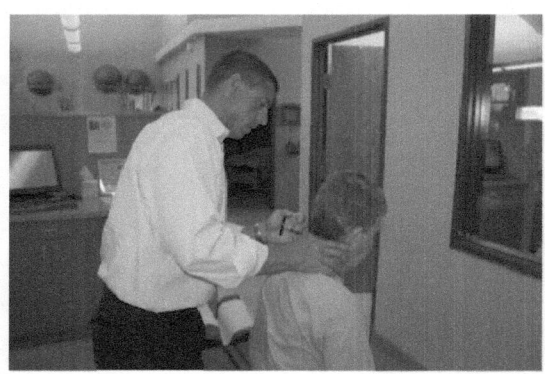

Your central nervous system regulates and controls everything in your body, from thinking to blood flow to sight and touch. You can go weeks without food, days without water, and minutes without air, but you cannot survive for one second without a functioning nervous system. You could eat right and exercise until you are blue in the face, but if your body does not produce the right hormones for digestion and growth, these practices do no good. Making sure that your nervous system is optimal is just as important as eating right and exercising. Every nerve in your body exits through your spine, and even minimal amounts of pressure on the nerve keep it from functioning at 100 percent. Subluxation, the major cause of pressure on nerves, is a movement deficiency of our bones that will lay down fibrous adhesions and scar tissue around the joint, choking off the nerves. In a Scandinavian study, gastroenterologists found that 72 percent of patients with abdominal pain, IBS, and heartburn had subluxations of the spine in the area that supplies nerves to the abdomen. (*Journal of Gastroenterology*, 1990, Dec 25, 12)

Scoliosis, or curvature of the spine, has been found to affect the digestive system. According to the Journal of the American Medical Association, scoliosis of even a slight degree involving segments corresponding to the stomach and duodenum seems to play a part in 90 percent of peptic ulcers. (Medical Journal Abstracts, *JAMA,* Nov. 15, 1958)

When you get adjusted to remove subluxation, your body moves back toward health. The adjustment creates movement in the spine, impacts the brain and nervous system to produce hormones needed for growth and repair, and inhibits cortisol and catecholamines, the two major players in sickness and disease.

You are what you can process, not just what you eat!

Chapter 4

The Realities of Exercise

Many health experts say that the role of exercise in any weight loss program is simply to burn calories. Their reasoning is that if you burn more calories than you eat, you will lose weight. But if you don't burn as many calories as you eat, you will gain weight. Simple enough, right?

Although this reasoning makes sense and is true for the most part, it fails to capture the full range of physiological and mental changes that occur when you exercise. Losing weight is just the tip of the iceberg. Exercise improves your blood pressure, decreases your cholesterol level, decreases your risk of heart disease and cancer, improves your mental functioning, elevates your mood, improves your quality of sleep, keeps you from losing muscle during weight loss, decreases your emotional and physical response to stress, and improves your flexibility and balance. In this chapter you will learn how exercise can improve your life and how to integrate exercise into your busy schedule with *the 6 Steps to Exercise for Health and Wellness.*[TM]

Exercise, the Fountain of Youth

In their landmark book entitled *Biomarkers*, medical researchers Dr. William Evans and Dr. Irwin Rosenberg identified ten characteristic changes that occur as we age.

- Loss of muscle mass
- Decrease in strength
- Decrease in basal metabolic rate
- Decrease in aerobic capacity
- Increase in blood pressure
- Loss of normal insulin action

- Decrease in circulating HDL to total cholesterol ratio

- Loss of bone density

- Decreased ability to control body temperature.

Each of these measures of physical health contributes to the decline that accompanies aging, but Evans and Rosenberg discovered that exercise is effective in reversing every one of these markers.

 When we think of exercise and fitness, many of us from the baby boom generation think of Jack LaLanne, who hosted his own television exercise show for thirty-four years beginning in the early 1960s. He was a sugar-addicted weakling for most of his youth, and at one point the family doctor told his parents he might not have long to live due to his ill health. That changed the day Jack decided he was going to quit eating junk food and begin exercising.

Over the ensuing years, Jack accomplished what many would consider impossible feats, such as performing 1,033 push-ups in twenty-three minutes on national TV at the age of forty-two, and on his seventieth birthday, swimming 1.5 miles in San Francisco Bay while pulling seventy boats! Jack died January 23, 2011 at ninety-six years young. He still exercised for two hours every day just months before he died—one hour of swimming and one hour of weightlifting.

You don't have to exercise as intensely as Jack LaLanne in order to enjoy the benefits of exercise. But you do have to get up and get your body moving at least five or six days a week; preferably doing a combination of aerobic exercise and strength training.

The Benefits of Aerobic Exercise

Aerobic means "using oxygen," and aerobic exercises are those that utilize oxygen during the activity. Aerobic exercise trains the body to utilize oxygen more efficiently and improves your overall cardiovascular fitness.

Aerobic activities are those that are performed for an extended period of time at a low intensity. Examples of aerobic activities are biking, aerobic walking, swimming, jogging, in-line skating, aerobic dance, cross-country skiing, and using an elliptical trainer. The benefits of aerobic activity include

- Improved breathing

- Increased energy throughout the day

- Improved heart health and cardiac output

- Decreased blood pressure

- Decreased serum cholesterol

- Decreased stress

- More restful sleep

- Improved mood and mental functioning

- Improved digestion and bowel function

For maximum benefit, you should engage in at least twenty minutes of aerobic activity five or six days a week. If you can do more than this, great! For those who have not engaged in regular activity for a while, even twenty minutes a day will be an accomplishment. During aerobic exercise, you should be able to carry on a conversation without feeling winded. If you are breathing too hard to talk easily, you should ease up a bit. As you become healthier, you will be able to exercise more intensely without feeling short of breath.

This brings up another important point—a concept called the *overload principle,* which states that in order to benefit from physical activity, the intensity has to be greater than your body is used to. Only if you push your body a little—overload it—will it respond by growing stronger.

Staying in the Aerobic Training Zone

The easiest way to tell if you're exercising intensely enough is to measure your heart rate. The following table of heart rate ranges will tell you how to find out if you are exercising in the aerobic zone or outside it in either direction. Find your age on the table and look for your aerobic heart rate range. Remember these numbers, and at least once during your daily aerobic activity, take your pulse. If you are below this range, you will need to step up the pace a bit. If you are too high, slow down a little.

Your target aerobic heart rate range is calculated by subtracting your age from 220 to find the maximum heart rate for your age. Any exercise that keeps your heart rate in the range of 65 to 75 percent of your maximum is optimal to burn fat and improve cardiovascular fitness.

The Benefits of Strength Training

Strength training differs from aerobic training in three important ways. First, strength training involves activities that are more intense and much shorter in duration than aerobic activity, such as push-ups or sit-ups. Second, while aerobic training primarily improves the health of your cardiovascular system, strength training primarily improves the health of your muscles, bones, and joints. Third, while aerobic activity should be performed almost every day for maximum benefit, you only need to engage in strength training two to three times a week. The benefits of strength training include:

- Increased muscle and bone strength
- Improved muscle tone and body shape
- Improved hormone function
- Improved mood and mental functioning
- Decreased serum cholesterol
- Decreased stress
- More restful sleep
- Increased metabolism
- Prevention of muscle loss associated with dieting

To benefit from strength training, it is not necessary to spend long, grueling hours in the gym every day. In fact, you can experience a significant improvement in your strength and muscle tone by weightlifting for an hour only two or three times a week. The key to successful strength training is not the amount of time you spend, it's the intensity. The harder you work your muscles during your strength workouts, the quicker you will see improvements.

How Your Muscles Respond to Strength Training

Your muscles do whatever you tell them to do. Strength training is simply a way to tell your muscles that you want them to get stronger. If you tell them often enough, they will listen.

Each time you exercise your muscles hard, your body goes to work to build more muscle. As long as you slowly increase the weight you use week after week, your muscles will continue to strengthen. Most people begin to see a difference in strength and appearance after a few weeks of strength training.

There are a lot of misconceptions about strength training, such as the myths that if you stop lifting weights, the muscle you gained turns to fat, and that strength training is not good for women. Neither of these is true.

Exercise Tips for the Overweight

Heavy people face special challenges in trying to be active. You may not be able to bend or move in the same way other people can. It may be hard to find clothes and equipment for exercising, and you may feel self-conscious being physically active around other people. Facing these challenges is difficult, but it can be done.

When starting your exercise program, it is important to go easy on yourself. If you cannot do an activity, don't feel bad about it. Just feel good about what you can do and avoid negative self-talk. An overweight woman once told me that if other people talked to her the way she talks to herself, she would punch them. Focus on the positive and you will improve your chances of success.

It is important to start slowly. Your body needs time to get used to your new activity. Be sure to spend some time warming up to get your body ready for action. Shrug your shoulders, tap your toes, swing your arms, or march in place. You should spend a few minutes warming up for any physical activity, even walking. Walk slowly for the first few minutes. Afterward, spend some time cooling down. Slow down little by little; walk more slowly or stretch for a few minutes. Cooling down protects your heart, relaxes your muscles, and keeps you from getting hurt.

Overweight people can do most of the physical activities in this book. Here are some examples.

- Weight-bearing aerobic activities, like walking or using the elliptical machine, which involve lifting or pushing your own body weight.

- Non-weight-bearing aerobic activities, like swimming, water workouts, and the exercise bicycle. These activities put less stress on your joints because you don't have to lift or push your own weight. If your feet or joints hurt when you stand, non-weight-bearing activities may be best for you.

- Most strength training exercises, such as lifting weights, can be performed while seated or lying down. Strength training is the quickest and safest way for heavy people to increase metabolism, improve strength, and burn calories while minimizing stress on the feet, knees, and ankles.

Be sure to pay attention to your body. If you are heavy, your joints carry more weight than those who are leaner. If your feet, knees, or back begins to hurt from weight-bearing exercise, start by doing non-weight-bearing exercise and slowly work up to weight-bearing exercise. Listening to your body will help you avoid injuries that can set you back.

If you are not active now, start slowly. Try to walk four minutes per day during the first week and eight minutes per day the second week. Walk eight minutes at a time until you feel comfortable, and then increase your walks to twelve minutes. Lengthen each walk by four minutes each week until you reach twenty minutes per day. At that point, you can work on quickening your pace to get your walking into the aerobic range. Please be patient! You will be much better off in the long run if you start slowly and build slowly.

You can do many activities in your home, but there are advantages to exercising in health clubs, including access to exercise equipment that can support heavier people and being with other people who can provide inspiration and support. You may feel self-conscious exercising around other people, but keep in mind that you have just as much right to be healthy and active as anyone. Even though you may feel as if other people are judging you, they are not. Many of the people who exercise regularly might be very supportive and they can be a great source of information.

6 Steps to Exercise for Health and Wellness

Step 1. Exercise at the same time every day.

The people who are the most successful at consistently exercising work out at the same
time every day. More than 90 percent of the things you do every day are out of habit. The time you get up each morning, when you leave your house, the route you take to work, whether you take the stairs or elevator, what you do when you get home—all of these are done habitually. Successful people realize that you are your habits, and if you can change your habits you can influence your health potential. We advise all our patients to begin setting their alarms thirty minutes earlier. When this becomes a habit, you can do this even if you are away on a trip or on vacation. There are a million reasons that can come up in a day for you to not exercise, but there is nothing scheduled for thirty minutes before you have to get up. Also, working out in the morning allows the metabolism to be elevated throughout the day. However, if this is not possible, working out on your lunch hour is the next best option.

Step 2. Make changes one at a time.

When exercise and diet programs fail, it is usually due to the element of pain, whether physical, emotional, or both. It has been said that people will do anything to avoid pain and seek pleasure. Starting to exercise and keeping it up is difficult enough without having to change your diet at the same time. We have noted that it is a good idea to add things when we take things away. By adding exercise in the morning, even if you maintain your current eating patterns, you are beginning to move toward homeostasis. This is more likely to become a habit if we don't change too many things at once. After thirty to sixty days we can make another change in diet or exercise, and then another change can occur after a new habit is established.

Step 3. Hold yourself accountable.

Starting a new program can be a little scary. If you have difficulty designing an exercise plan, staying motivated, or staying on schedule, then get help. You might get a workout partner or start a walking club at work. As a pair or group, you not only help and support one another, you are also accountable to one another if you choose to be. If someone is coming over to work out with you or if you are meeting others to exercise as a group, you are less likely to skip your workout or hit the snooze button in the morning. You might also hire a personal trainer or chiropractic wellness coach to guide you through the process.

Step 4. Take a before picture.

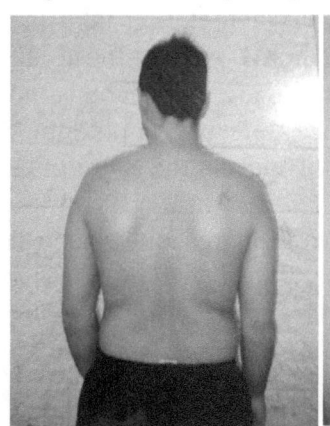 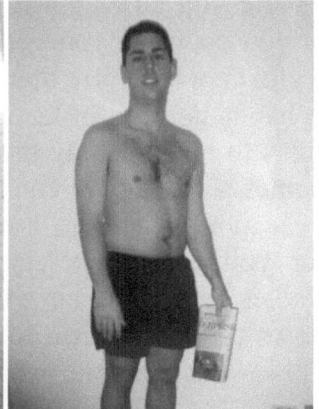 

There is nothing more humbling than seeing a picture of yourself from the front, side, and back, but it does give you a baseline for your future progress. When you get frustrated or angry, or when you feel stuck, you can look at the picture of where you started. Scales don't tell the story; a picture is worth a thousand words. After you take your photograph, make sure you get your body fat tested. There are several ways to do this. but having it done by a professional with calipers is the most accurate.

Step 5. Plan to alternate aerobic and nonaerobic exercise.

Any successful workout program will have both aerobic and anaerobic exercise. When you set up your workout schedule, you should alternate strength training with aerobic exercise. You might work out with weights on Sunday, Tuesday, and Thursday, and run, swim, or bicycle on Monday, Wednesday, and Saturday, or any other combination of alternating days. Doing this, your body has a chance to recover from one type of exercise while you do the other. The remaining day can be a rest from both.

Step 6. Don't do a lifestyle change.

Many people set themselves up for failure from the beginning. Thinking that on Monday morning you are going to start exercising and eating right as a part of a new lifestyle never works. Most people feel pain and miss the pleasure they had from food. Then they decide they could never go the rest of their life like this. We advise all our patients that until it becomes a habit, have an ending day. Use a date such as your birthday, a wedding, or reunion, about three to six months off. If there is an ending date and a goal, a person can go through just about anything. In our office we break it down even further, to every six days. Sunday is a recommended day off, but any day can be considered a free day. On your free day, not only do you not exercise, but you can eat whatever you want. That's right, I said whatever you want and however much you want. This way you only need to make it through six days. You can do this as long as you want, but when you start to feel better, you might want to keep your new routine seven days a week.

Exercise for Health and Wellness

Health and Wellness Exercise Plan
Weight Training 3 times per week
5 minute warm up: light weights
Rotator cuff: 15–20 reps

Bench press 15–20 reps

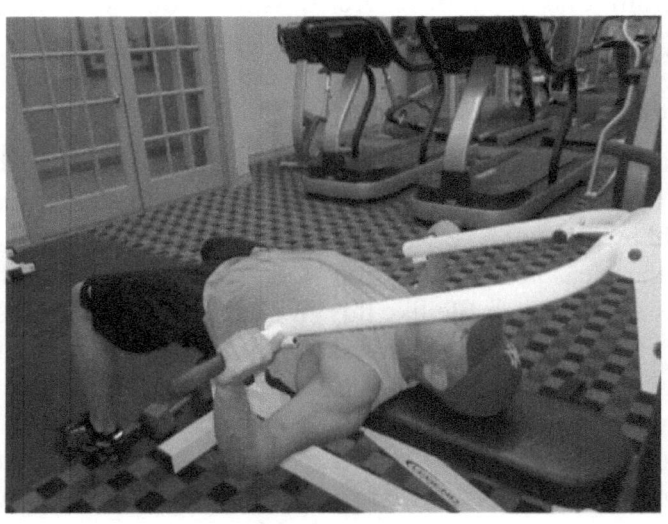

Lat pull-downs 15–20 reps

Super set the exercises then take a rest.

Do flat bench then go over and do chin-ups, then take a 30 second rest and start over again. You also can mix up the exercises so you could do fly's with lat pull–downs—whatever works for you—but make sure you do the exercises listed below.

This is a total body workout so each time you lift weights pick one superset per muscle group. Make sure that each day you do a different superset for each group!

Start off at ten reps then increase reps and go to failure!

Chest and back

Flat bench then chin-ups 3 sets/10 reps

Incline bench then lat pull-downs 3 sets/10 reps

Fly's then seated rows 3 sets/10 reps

Anterior shoulder then posterior shoulder

Ball push-up 3 sets/10 reps

Triceps and biceps

 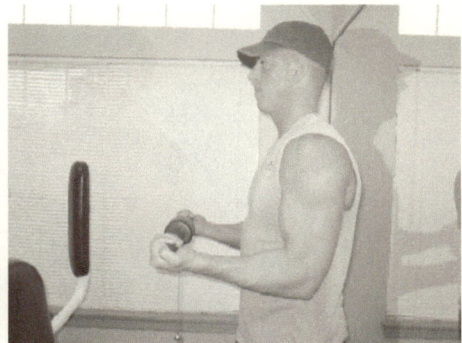

Dips then straight bar curls3 sets/10 reps

Triceps kick backs then incline curls3 sets/10 reps

Skull crushers then preacher curls3 sets/10 reps

Quads/Gluts and hamstrings squats3sets/10 reps

Lunges (forward/back/side) 3 sets/10 reps

Hamstrings 3 sets/10 reps

Cardio days

Run or bike for 60 minutes

Add interval training 2 times per week

Sprint, jog (3 min on, 3 min off)

Range of Motion for Health and Wellness

It is essential to your health to have full range of motion in your spine. The following exercises are simple and easy to do and should be done daily for maximum effect.

Core Stretching for Health and Wellness

The best way to prevent injury is to have strong, flexible muscles and joints that resist strain and injury. In some simple cases of back pain, exercises can relieve pain. Remember, never do any exercise that causes increased pain.

Stretching Workouts for the Beginner

Press Up: Sphinx Position

Start by lying on your stomach. Begin to raise your upper body slowly, while keeping your pelvis flat to the floor. Try to create an arch in your low back. Go up only as far as you can without discomfort. Work up to the position shown here, known in yoga as the Sphinx position.

Superman

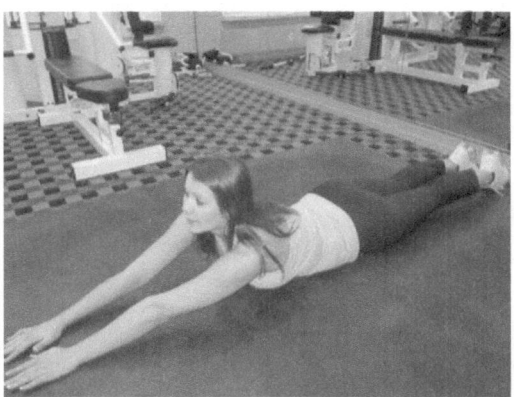

Start by lying on your stomach face down. Raise your shoulders and hold yourself up with your arms extended in front of you. Hold for 30 seconds, then return to starting position. Repeat ten times.

Standing Back Extension

This exercise can be done at work or any other place where doing a press-up on the floor is practical. Start with hands on low back. Slowly arch backward as far as you can without discomfort. Hold for only three seconds and return to starting position. Repeat five times.

The "Dog"

Start on all fours. Create an arch in your low back by lowering your abdomen toward the ground, while at the same time raising your head. Hold for 30 seconds. Go back to starting position. Repeat 20 times.

Single Knee to Chest

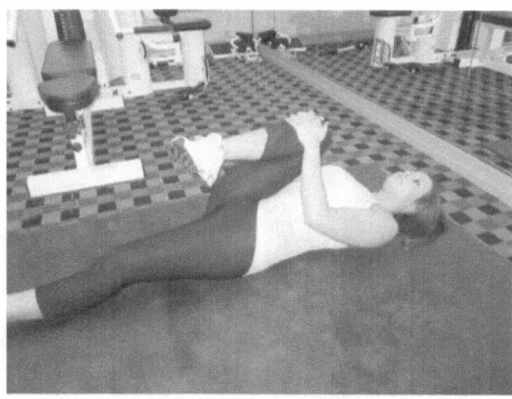

Start with both legs together, heels flat on the ground. Raise your right knee upward with both hands and pull it toward your chest. Hold for 30 seconds and return to starting position. Repeat with other leg. Do ten repetitions with each leg, alternating between right and left.

Piriformis Stretch

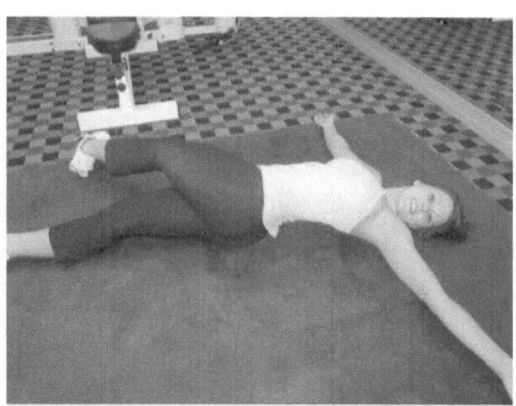

Lie down with your right knee up and both arms stretching outward at 45 degree angles away from your body. Slowly let your right knee fall across your body as far as you can, keeping your shoulders as flat as possible. Hold for 30 seconds. Return to starting position. Do the same on the other side. Hold for 30 seconds. Return to starting position. Do the ten repetitions, alternating between right and left.

Piriformis Standing

This exercise can be done at work, or during recreational activities such as golf, because it can relieve back pain without requiring you to lie down.. To help you maintain your balance, you can lean against a wall, fence, or tree. Raise your knee in front of you and slowly swing it across your body. Hold for ten seconds and repeat, alternating knees

Body Flexion

Start on your knees with your hands across your abdomen. Lift your upper body and curl it forward very slowly, keeping your head off the ground. Hold for thirty seconds and then lower your body slowly back down. Repeat several times.

Body Flexion And Stretch

Start on your knees. Slowly lean forward and let your hands stretch out and forward. Be sure to keep your head off the ground. Hold for 30 seconds. Repeat several times.

Bend Over

Standing straight, cross your arms across your chest. Slowly bend over, a bit at a time, allowing the weight of your upper body to stretch your back. Relax as your back and the backs of your legs begin to stretch. Hold for ten seconds and returned to a standing position very slowly. Repeat several times.

Stretch to Foot

Start in a sitting position with legs extended and feet together. With your hands flat against the ground, slowly extend forward as far as you can comfortably. Hold for 30 seconds and relax. Repeat ten times.

The Plow

The Plow is an advanced yoga position that should be attempted only after you are pain-free and have mastered simple back stretches both forward and back. Lie on your back and slowly raise both legs. Keep going as far as you can without discomfort with the goal of reaching the Plow position, using your outstretched arms to balance. Hold for 30 seconds and slowly return to starting position. Repeat several times.

Press Up

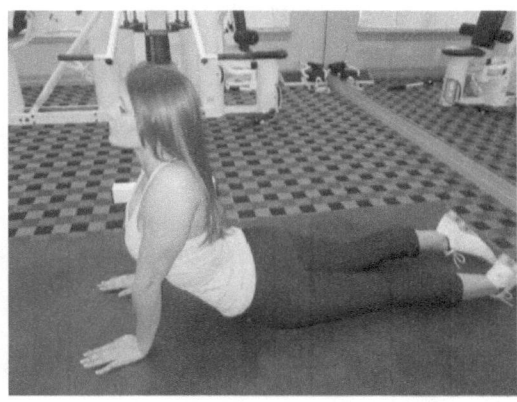

Start by lying on your stomach. Raise your upper body slowly while keeping your pelvis flat to the floor, arching your back and keeping your arms straight. Go up only as far as you can without discomfort, with the goal of reaching the Sphinx position.

Neck Exercises

Here are some simple stretching exercises to relieve neck pain, including the muscle strain that comes from sitting at a desk all day. You can do these exercises in your office chair as often as necessary to relieve neck strain. For a more detailed look at this please refer to my previous book- *Stay Fit While You Sit*.

Neck And Glide Extension

NECK GLIDE (middle photo): Start with neck straight. Slowly slide your chin forward. Hold for five seconds and return to starting position. Do ten times. NECK EXTENSION (right photo): Without arching your back, slowly move your head backward so you are looking up. Hold for five seconds. Return to starting position (far left photo).

Neck Rotation

Start by looking straight ahead. Slowly turn your head to the left. Hold for ten seconds, then return to starting position. Slowly turn your head to the other side. Hold for ten seconds. Return to starting position. Do ten repetitions. This exercise can be done at your desk as often as necessary to relieve the stress of extended periods working on a computer.

Neck Side Extension

Start by looking straight ahead. Slowly lean your head to the left. Hold for five seconds, then return to starting position. Slowly lean your head to the other side. Hold for five seconds. Return to starting position. Do ten repetitions. This exercise can be done at your desk as often as necessary to relieve the stress of extended periods working on a computer

Neck Stretch

Looking straight ahead, slowly raise both shoulders andold for five seconds, then return to starting position. Do ten repetitions. This exercise can be done at your desk as often as necessary to relieve the stress of extended periods working on a computer

NOTE: We recognize that people will diagnose and treat themselves. We have provided this medical information to make you more knowledgeable about nonsurgical aspects of care, the role of exercise in your long-term recovery, and injury prevention. In some cases, exercise may be inappropriate. Remember that if you diagnose or treat yourself, you assume the responsibility for your actions. You should never do any exercise that causes increased pain. You should never do any exercise that places body weight on a weakened or injured limb or back. A special thank you goes out to Prizm Development for using many of the exercises.

Posture for Health and Wellness

Posture exercises can help loosen the neck and shoulder muscles that hold so much of our tension, especially when we sit all day at a desk and hunch forward over a computer. When you stand straight, your ears should be directly over the middle of your shoulders and your shoulders should line up with your hips. Your hip should be in alignment with your knee and your knee in alignment with your ankle. Ask someone to check your posture, and you might want to do the same for them. These postural exercises will help you stand straighter, feel better, and look more at ease in your body. You can remind yourself to do the exercises every day by remembering that *Y*ou *W*ant *T*o *L*ive a healthier life.

Y Stretch. Place your arms above your head with your thumbs out and back and pull your shoulder blades back as far as you can. Hold as long as you can and repeat ten times.

W Stretch. Place your arms out at a 45-degree angle and pull your shoulder blades back as far as you can. Hold as long as you can and repeat ten times.

T Stretch. Place your arms at a 90-degree angle, palms down, and pull your shoulder blades back as far as you can. Hold as long as possible and repeat 10 times.

L Stretch. Place your arms at your sides, elbows out at a 90-degree angle, and pull your shoulder blades back as far as you can. Hold as long as possible and repeat ten times.

These exercises are a little more advanced and should be done only when they feel comfortable. Stop stretching if you feel any pain.

Back on Wall Stretch. Place your feet against the wall and your buttocks. Bring your arms up to 90 degrees, keeping your chin tucked in and your back against the wall. Pull your shoulder blades back and hold as long as possible. Repeat ten times.

Doorway Chest Stretch. Stand in a doorway with your arms up at 90 degrees and lean forward, stretching your chest. Hold as long as possible and repeat ten times.

Chapter 5

Stress Reduction

Stress exists everywhere—we cannot avoid it. Worrying about job security, being overworked, driving in rush-hour traffic, arguing with your spouse, —all these create stress. According to a recent survey by the American Psychological Association, 54 percent of Americans are concerned about the level of stress in their everyday lives, and two-thirds of Americans say they are likely to seek help for stress. You are not alone, but there are things you can do to minimize stress and manage the stress that's unavoidable.

What is Stress?

Stress is a force that causes change in your life. It's a feeling that's created when you react to particular events. It's the body's way of rising to the challenge and preparing to meet a tough situation with focus, strength, stamina, and heightened alertness. The events that provoke stress are called stressors, and they can be anything from the threat of physical harm to preparing for a public speech to getting ready for a date.

We are continually adapting and responding to stress on all levels at all times, without even realizing it. The pupils in our eyes are continuously adapting to changes in the brightness of the room. The body's pores and muscle tone are continually adapting to changes in the temperature. Even the gravity that keeps us on the earth is a constant stress on the body. But stress is not judged by the body as good or bad. It is used by the body, and it elicits a reaction that the body interprets to be to its best advantage. In other words, your body responds in the most appropriate way to the stressors it is given.

Stress can be divided into three types: physical, chemical, and emotional.

Physical Stress

Gravity is always at play. We are constantly adapting to the physical demands that gravity places on us. In fact, if I ask you to sit up straight and you find yourself having to make corrections in your posture, such as lifting your head up or pulling your shoulders back, these changes indicate that your body is trying to adapt you to the effects of gravity and the stress it represents.

Physical stress is probably the most serious if left undetected or uncorrected. It can lead to postural problems, subluxation, or Forward Head Posture (FHP), all of which can lead to health problems. Your posture changes instantaneously, but your symptoms may not show up until after years of problems and physical stress. Your Doctor of Chiropractic, as a spinal specialist, can detect stress by studying your posture and what it says about your health. If you pay no attention to your posture or if you say "Oh, that's just my posture. I know it's bad but I can't do anything about it," you are ignoring the plan for your body to use posture as a tool to detect stress-related problems early.

Renee Calliet, M.D., found that a forward head position of three to five inches adds hundreds of pounds of tension stress to the cervical spine. Any spinal tissue that is subject to a sustained load over a period of time will begin to remodel out of proper position.

Dr. Roger Sperry, who in 1980 won a Nobel Prize for brain research, has found that more than 90 percent of the energy output of the brain is used "in relating the physical body [to] its gravitational field. The more mechanically distorted a person is, the less energy [is] available for thinking, metabolism, and healing."

Chemical Stress

We are constantly interfering with our body's ability to heal and to maintain health and wellness by poor nutrition, unnecessary medications, and toxins. Food is your energy. It enables you to move and to think. The quality of the food you consume will have a direct effect on the amount of  energy you have and on your ability to concentrate.

There is much truth to the saying "You are what you eat." You should eat a variety of whole foods throughout the day. This is such an important aspect of stress I have devoted two chapters in this book on the topic. For more information, please read the chapters on nutrition and toxins.

Emotional Stress

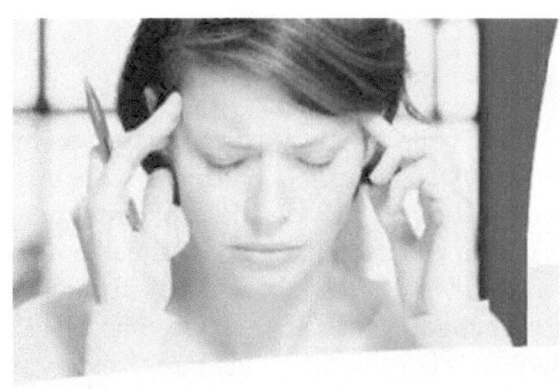 Emotional stress occurs when you worry about money, a loved one's illness, or retirement, or when you experience an emotionally devastating event, such as the death of a spouse or losing your job. We have helped hundreds of people deal with hundreds of stressful problems by pointing out that emotional stress does not exist. "What do you mean?" our patients ask. "What about my increased blood pressure, increased heart rate, sweating, dry mouth, and difficulty breathing?"

These sensations are not from stress. They are from an emotion that starts in a pea-sized part of the brain called the *amygdala*—the age-old emotion of fear. Children never say they're stressed about the thunder or having anxiety about the boogieman; they're scared and they say so. These fears continue into adulthood in the form of fear of financial catastrophe, of losing people we love, of losing our jobs, and in general, fear of the unknown. The stresses we face in life trigger our fears.

The Fear Response

Often referred to as the *fight-or-flight reaction*, the stress/fear response occurs automatically when you feel threatened. Your body's fight-or-flight reaction has strong biological roots in the need for self-preservation. Early humans had to fight aggressors and run from predators, and this response was important to survival. Now, however, instead of protecting us, the fight-or-flight reaction can have the opposite effect. If you are constantly stressed or afraid, you may be more vulnerable to life-threatening health problems.

Any sort of change in life can make you feel fear, even good change. It's not just the change or event itself, but how you react to it that matters. What may be stressful is different for each person. For example, one person may not feel stressed by retiring from work, while another may be afraid of the changes in life that retirement represents. Successful people realize that the bigger the challenge that shows up in your life, the more fear will come with it.

How Fear Affects Your Body

When you experience fear, your pituitary gland responds by releasing a hormone called *adrenocorticotropic hormone* (ACTH). When the pituitary sends out a burst of ACTH, it's like an alarm going off deep inside your brain that tells your adrenal glands, situated atop your kidneys, to release a flood of stress hormones into your bloodstream, including cortisol and adrenaline. These hormones cause a whole series of physiological changes in your body, such as increasing your heart rate and blood pressure, shutting down your digestive system, and altering your immune system. Once the perceived threat is gone, the levels of cortisol and adrenaline decline, and your heart rate, blood pressure, and other body functions return to normal.

Let me share with you a story from James Chestnut, D.C., as an example of how stress affects the physiology of your body.

Let's say I live 40,000 years ago, at the time of our hunter-gatherer ancestors. At dawn I decide to go down to a little brook to get a drink of water. There is a mist coming off the water, and as I look across, I see a fawn putting its lips to the water and having a drink too. Then all of a sudden a saber-tooth tiger suddenly appears! What does my body do? What does my physiology do? Does it say, "Well, maybe it's an old tiger, and perhaps it doesn't have any teeth left. I'm just going to chill out and check." Or does my body prepare me to fight or flee, just in case? It's a sure thing that I'd better be ready to do battle or get out of there!

So what is the very first thing that happens to your physiology in that scenario? Your nervous system detects something in your environment. All that you are is based on the environmental exposure that you've had. The things you have been repeatedly exposed to and those that have caused a high emotional charge are the ones that have had the most dramatic impact on your life. You don't forget the events that happened when you were under a lot of stress.

For example, if you were vacationing in Florida and an alligator popped out of the bushes and snapped at you, you would never again forget to check nearby bushes. Because this was an emotionally charged event, it caused hormones to be released in your amygdala and triggered a fight-or-flight response, causing you to be careful around bushes from then on.

In fight-or-flight, what happens physiologically? Your hypothalamus tells the cells of your sympathetic nervous system, the part of the nervous system that triggers the response, to increase your heart rate, increase your respiratory rate, increase your blood pressure, and increase your cardiac output. Support for growth and repair shuts down, and support for breakdown gears up. Breakdown doesn't sound smart, but it is smart if it can save you from the tiger or the alligator. It's smart if it can help you survive to change your environment. The only reason you have a fight-or-flight response is so you can survive to get into a more hospitable environment, because it's the environment that determines your health.

Next, your sympathetic nervous system signals to your adrenal glands to start producing epinephrine and adrenaline. You produce epinephrine and adrenaline in the amygdala, your stress and anxiety center, as well, because your brain has to get ready for fight or flight too. Your brain shuts off the parts for short-term memory and learning, because you aren't going to try to remember your grocery list while a tiger or an alligator is nipping at your leg. You go into an instinctual part of your brain in order to survive.

Stress causes you to sharpen your signal detection system at the expense of thought, and that's why people who are always stressed out can't concentrate. They become more sensitive both emotionally and physically, because that is the body's instinctual response to stress.

So the adrenaline and epinephrine are released, and their job is to liberate free fatty acids to be used as your rocket fuel. They stop your insulin receptors from absorbing sugar so that it can remain in your bloodstream for energy. Stress hormones also promote the release of clotting factors and platelets to aid in healing if you are injured.

Your body will release cortisol to promote production of sugar for energy, break down proteins used to produce sugar, and inhibit your immune system. I know it doesn't sound smart to turn off the immune system while you fight or flee, but why would you need it? After all, white blood cells won't save you from a predator. You might need your immune system if you survive the attack, but it is of no use to you during fight or flight. The immune response is metabolically expensive; it uses up a lot of the body's resources. That's why you feel tired all the time when you have a cold or the flu.

Under stress you crave fats and sugars, so you gain weight. Back in the day, you would get your fats from a wildebeest or some nuts and seeds, and sugar would come from fruits and vegetables. Now, though, we can fill ourselves with sweets, starches, and other comfort foods when we feel stressed. Refined sugars will cause your blood glucose levels to increase, which will cause your pancreas to secrete more insulin. Insulin is one of the most powerful stimulators of the stress response, and now we have a vicious cycle going that leads to a chronic stress reaction.

The body reacts to stress by increasing the level or intensity of several bodily functions, including the following.

- Heart rate

- Blood pressure

- Blood sugar

- Blood fats and cholesterol

- Insulin resistance

- Insulin levels

And how does that sound as far as being predisposed to obesity, type II diabetes, heart disease, and stroke? We are pretty much right on track for those now, aren't we? With our immune systems suppressed, we are a lock for more colds, flu, and cancers. You can see that stress is related to every chronic illness, so even though the stress response can save your life in a crisis, it can ultimately lead to your demise.

Long-term stressful situations can produce lasting, low-level stress that wears down the body. The nervous system senses continued pressure and remains slightly activated, continuing to pump out extra stress hormones. This can deplete the body's reserves, leave you feeling fatigued or overwhelmed, weaken your immune system, and reduce health and well-being drastically over time.

If stressful situations pile up one after another, your body has no chance to recover. This long-term activation of the stress/fear-response system can disrupt almost all your body's processes. There are certain systems that can be counted on to react badly to stress.

- **Digestive system.** Stomachaches and diarrhea are very common when you're stressed because stress hormones slow the release of stomach acid and the emptying of the stomach. The same hormones also stimulate the colon and speed the passage of its contents.

- **Immune system.** Chronic stress tends to dampen your immune system, making you more susceptible to colds and other infections. Typically, your immune system responds to infection by releasing substances that cause inflammation. Chronic systemic inflammation contributes to the development of many degenerative diseases.

- **Nervous system.** Stress has been linked with depression, anxiety, panic attacks, and dementia. Over time, the chronic release of cortisol can cause damage to several structures in the brain. Excessive amounts of cortisol can also cause sleep disturbances and a loss of sex drive.

- **Cardiovascular system.** Stress causes an increase in heart rate and blood pressure and increases the risk of heart attack and stroke.

Responses to specific stressors vary widely. Some people are naturally laid-back about almost everything, while others react strongly at the slightest hint of stress. Any of the following conditions might be a sign that you are suffering from stress.

- Anxiety
- Insomnia
- Back pain
- Relationship problems
- Constipation or diarrhea
- Shortness of breath
- Depression
- Stiff neck
- Fatigue
- Upset stomach
- Weight gain or loss

Reducing the effects of fear

What can you do to deal with the effects of stress, or better yet, avoid them in the first place? After decades of research, it is clear that the negative effects associated with stress are real. Although you may not always be able to avoid stressful situations, there are a number of things you can do to reduce their effects. You can learn how to manage the stress that comes along with any challenge. Stress management skills work best when they are used regularly and not just when the pressure's on. Learning stress reduction methods and mastering them when you are calm will help you get through the times when your biggest challenges are upon you.

The first is to identify the fear. Find out the possible consequences of what you're afraid of. Many times it isn't as bad as you think. Think of FEAR as *False Evidence Appearing Real*. If you discuss your fears with a friend or colleague, a solution might well present itself. People are more likely to listen and help someone when they talk about fear than when they say they're stressed out.

The second way to reduce fear is to relax. Learning to relax doesn't have to be difficult. Here are some simple techniques to help get you started on your way to health and wellness.

Relaxed breathing

Have you ever noticed how you breathe when you're stressed or afraid? You *stop* breathing. We hold our breath and decrease our blood oxygen levels, immediately irritating our body and producing toxic chemicals. Either that or we hyperventilate. Stress can cause rapid, shallow breathing, which sustains other aspects of the stress response, such as rapid heart rate and perspiration. If you can get control of your breathing, the spiraling effects of acute stress and fear will automatically become less intense. Relaxed breathing, also called *diaphragmatic breathing*, will help you, especially if you practice this basic technique twice a day every day and whenever you feel tense. This exercise combines good posture, which affects your brain and spinal cord, with enhanced oxygen levels in the blood.

Inhale. With your mouth closed and your shoulders relaxed, inhale as slowly and deeply as you can to the count of six. As you do that, push your stomach out. Allow the air to fill your diaphragm.

Hold. Keep the air in your lungs as you slowly count to four.

Exhale. Release the air through your mouth as you slowly count to six.

Repeat. Complete the inhale-hold-exhale cycle three to five times.

Progressive muscle relaxation

The goal of progressive muscle relaxation is to reduce the tension in your muscles. First, find a quiet place where you'll be free from interruption. Loosen tight clothing and remove your glasses or contacts if you'd like. Do these exercises for about ten minutes at least once a day for maximum benefit. Tense each muscle group for at least five seconds and then relax for at least 30 seconds, and repeat before moving to the next muscle group.

- **Head:** Lift your eyebrows toward the ceiling, feeling the tension in your forehead and scalp. Relax. Repeat.

- **Face:** Squint your eyes tightly and wrinkle your nose and mouth, feeling the tension in the center of your face. Relax. Repeat.

- **Mouth and chin:** Clench your teeth and pull back the corners of your mouth toward your ears. Show your teeth like a snarling dog. Relax. Repeat.

- **Neck**: Gently touch your chin to your chest. Feel the pull in the back of your neck as it spreads into your head. Relax. Repeat.

- **Shoulders:** Pull your shoulders up toward your ears, feeling the tension in your shoulders, head, neck and upper back. Relax. Repeat.

- **Upper arms:** Pull your arms back and press your elbows in toward the sides of your body. Try not to tense your lower arms. Feel the tension in your arms, shoulders and into your back. Relax. Repeat.

- **Hands and arms:** Make a tight fist and pull up your wrists. Feel the tension in your hands, knuckles and lower arms. Relax. Repeat.

- **Chest, shoulders and upper back:** Pull your shoulders back as if you're trying to make your shoulder blades touch. Relax. Repeat.

- **Stomach:** Pull your stomach in toward your spine, tightening your abdominal muscles. Relax. Repeat.

- **Thighs**: Squeeze your knees together and lift your legs up off the chair or from wherever you're relaxing. Feel the tension in your thighs. Relax. Repeat.

- **Calves:** Raise your feet toward the ceiling while flexing them toward your body. Feel the tension in your calves. Relax. Repeat.

- **Feet:** Turn your feet inward and curl your toes up and out. Relax. Repeat.

Listen to soothing sounds

If you have about ten minutes and a quiet room, you can take a mental vacation almost any time. Consider two types of relaxation CDs or tapes to help you unwind, rest your mind, or take a visual journey to a peaceful place.

Spoken words. These CDs use spoken suggestions to guide your meditation, educate you on stress reduction, or take you on an imaginary visual journey to a peaceful place.

Soothing music or nature sounds. Music has the power to affect your thoughts and feelings. Soft, soothing music can help you relax and lower your stress level.

No one CD works for everyone, so try several CDs to find which works best for you. When possible, listen to samples in the store. Consider asking your friends or a trusted professional for recommendations.

Exercise

Exercise is a good way to deal with stress because it can relieve your pent-up energy and tension. It also helps you get in better shape, which makes you feel better overall. By getting physically active, you can decrease your levels of anxiety and stress and elevate your moods. Numerous studies have shown that people who begin exercise programs demonstrate a marked improvement in their ability to concentrate, are able to sleep better, suffer from fewer illnesses, suffer from less pain, and report a much higher quality of life than those who do not exercise. This is even true of people who do not begin an exercise program until they reach their 40s, 50s, 60s, or even 70s. If you want to feel better and improve your quality of life, get active!

Positive Mental Attitude: The Mind-Body Connection

Many of the things that happen in our day that cause us stress or trigger fear are allowed to occur because of our attitude. There is nothing anyone can say or do to you that can cause you to become stressed. A colleague of mine, Dr. Lazlo, used to say to me, "I stopped caring so much about what other people thought of me when I realized how seldom they did."

Our attitude is not a reflection of what others say or do to us, it is a reflection of what we say or do to ourselves. Having a good or bad attitude not only determines how you view the world but can play a huge part in your health. Aviad Haramati, M.D., Ph.D., the director of Complementary and Alternative Medicine at Georgetown University, says, "At this point there is enough basic research demonstrating the connection between mind and body that it is irresponsible to ignore such a significant issue in contemporary medical practice." It has been said that some people have willed themselves healthy beyond any medical explanation. I believe this can happen for two reasons.

We can change our physiology by what we think. For example, try to be depressed when you're thinking positive or fun thoughts and putting a smile on your face. Conversely, try to be excited with poor posture, slumped shoulders, negative thoughts, and a frown on your face. It can't be done, because our thoughts and physical actions determine our physiology.

As Tony Robbins will tell you, there was a research study done at UC-Berkeley where they did something that sounded insane. Traditionally if someone is depressed they go for the traditional psychiatric approach. Today we live with a philosophy that depression is merely a Prozac deficiency. We have a mindset that if things aren't working out, let's not resolve what's going on inside you, let's just drug you. It's called "better living through chemistry." Let's just drug you and keep on going. It's an interesting way we look at life today— we go for the quick hit. We want the instant solution, even if it's not the long-term solution that makes us healthier. Using that approach, we drug people.

At UC-Berkeley they tried something different: no drugs. Instead they came up with a simple approach. They took people that were clinically depressed, meaning they had achieved high levels of depression and had experienced it on a regular basis—and they had people come in for four weeks. They had them stand in front of a mirror and do one silly thing: grin from ear to ear. They made them smile for no good reason for twenty minutes, grinning so wide that they created crow's feet at the corners of their eyes. Also, they had to keep their shoulders back and breathe fully. What happened? Not a single person was able to remain depressed, including a woman who claimed that she was depressed even while she slept. At the end of these twenty days, many of them had no need for medication any more. All they did was change one part of their body!"

"Any stressor (toxicity or deficiency) that initiates the stress response will drive the mind-body away from homeostasis and toward adaptive physiology and illness," says Dr. James Chestnut.

The second reason I believe people can recover from sickness and disease beyond any medical explanation is faith.

Be Realistic

No one is perfect, so holding yourself to that high of a standard is unrealistic and can cause undue stress. Don't expect others to be perfect—it will cause stress in your life and put pressure on them. Be humble: when you feel overwhelmed, have a question, or just flat-out need some help, ask for it. *Ask and ye shall receive* is good advice.

Solve the Little Problems

By breaking down bigger problems into smaller tasks you are less likely to feel overwhelmed. Little problems always seem like they are much easier to solve. By taking on the smaller issues you build your inner confidence to tackle life's bigger challenges and you gain lots of momentum towards your ultimate goals.

Get a Good Night's Sleep

When you are asleep, your body is able to do repair and maintenance and recharge its batteries. Sleep keeps your body and mind in top shape, making you better equipped to deal with stressors.

The Power of Faith: How Ninth Grade Physics Changed My Life

I remember sitting in class learning about the Laws of Thermodynamics when something hit me like a lightning bolt. The teacher said "Energy is neither created nor destroyed; it is transferred." This is a law that can't be broken, like gravity. Even at that time, I knew the body was essentially energy. I knew life force surged when the brain sends electrical impulses down the spinal cord and out to the rest of the body. I knew the body and organs like the heart and lungs could be kept alive with technology, but when the brain stopped its electrical activity and generated no more energy, you are considered dead.

I began to think of what the teacher just said. If energy is neither created nor destroyed, just transferred, and we are essentially energy, then what happens when we die? This started a long list of questions to be considered along my spiritual walk. I have been blessed to have spiritual mentors and colleagues who have guided me to where I am today, like Coach Bill McCartney, Pastor Brian Thompson, and President of the Colorado Chiropractic Wellness Alliance Dr. Joe Arvay, all whom are extremely strong in their faith. Now I don't care what your faith is, but it is undeniable that 90 percent of the population believes in a God. Studies are now proving that spiritual health cannot be separated from holistic health. The studies of Harold Koenig, M.D. and David Larson, M.D. from Duke University have proven how spiritual belief increases longevity and protects against illness (Helm et al., 2000; Larson et al., 2000). Science has proven beyond a reasonable doubt that holistic health must include physical, chemical, emotional, and spiritual well-being.

Understanding the effects of energy and studying Deepak Chopra's work on quantum physics taught me that everything in the universe is nothing more than energy. Understanding how this energy works and how to use it to your advantage is the one of the great secrets to life. A good explanation can be seen in the DVD *The Secret* by Rhonda Byrne that sets forth the Law of Attraction.

The law of attraction simply states that like attracts like and what you think about you bring about through your thoughts. Dr. Joe Vitale in *The Attractor Factor* reminds us that our thoughts are nothing more than energy. When we are thinking about something, that signal has a frequency, and that frequency is drawing like frequencies back to you. Said another way, if you think you are healthy, you are right. If you think you are sick, you are also right. If you think you are a successful person or the opposite, you are right. If you think you can or you can't, you are always right. What you think about with vision and passion, you bring about. This is a universal law. If you see yourself healthy or living with abundance, through the law of attraction, you will attract it. The problem is that most people think about what they don't want and wonder why they are attracting it.

Bob Doyle said it well in an interview in *The Secret*.

> The law of attraction doesn't care whether you perceive something to be good or bad, or whether you don't want it or whether you do want it. It is just responding to your thoughts. So if you're looking at a mountain of debt, feeling terrible about it, that's the signal you're putting out in the universe. "I feel really bad because of all this debt I've got." You're just affirming it to yourself. You feel it on every level of your being. That's what you are going to get more of.

When you understand the law of attraction, you can see how profound this knowledge can be. If you are always complaining or talking about back pain or other health problems, what do you think you will get more of? That's right—what you are thinking and talking about will continue to be attracted to you because of this universal law.

Understanding this, you can think and talk about the solution or how you want to feel in the future. Through the law of attraction, the things, circumstances, and situations to bring this about will begin to present themselves.

Many times are in your current situation—physical, mental, emotional, financial—because you have attracted it into your life. There is so much more information on this topic that could be stated; however, this principle is underlying your current state of health. For more information, please read the aforementioned books or movie.

Be, Do, Have

Most people believe in a philosophy of *have, do, be*. For example, if I just *have* more money, then I could *do* what I need to *do* to *be* successful. Life rarely works this way. In fact, we all have more now than any generation of people has ever had, and we have more time to do the things that we want more than any generation before was able to do ... and yet we are not happy and satisfied. There is actually growing evidence that we have more emotional problems today than ever before.

We suffer from more tension, depression, fears, anxieties, etc. than ever before in our history. People are popping pills and taking tranquilizers as well as using drugs and alcohol, smoking cigarettes, and totally immersing themselves into television, the media, and the internet in an attempt to escape their reality. The simple truth is that if you reverse the order of the philosophy, it becomes be, do, have. If you are *being* happy and satisfied, you will automatically do what happy and satisfied people do and you will have the things that you want to have.

Let me break that down even further for you. Your *thoughts* create your *reality*, meaning that your life and how you feel are determined by you! You have the ultimate power to choose! Whatever thoughts that you have on a regular basis becomes your reality. So your unlimited power lies in your ability to control your thoughts. This means that however you perceive any given situation is how you are going to respond to it. For example, for most people driving in traffic, a reckless driver's cutting them off elicits an immediate stress response followed by road rage toward the reckless individual. We tend to show those emotions with hand gestures and yelling. We can stay in a stressed state for the duration of our trip or even the entire day. I used to be that guy who remained stressed for a day over a momentary incident. Now when someone cuts me off, I think to myself, "Yes! I avoided another accident!" Sounds silly, doesn't it? But it changes my state from being something negative that could affect my stress and my emotions all day, to something positive where I have just reinforced again that I am fortunate and I know I am going to have a great day.

So you see, your *thoughts* create your *reality*. Sadly, with the mainstream media the way it is, we have been bombarded our whole lives with problem after problem. If you don't believe me, just watch the morning or late night news, pick up the daily paper, or watch your TV shows. They are all inundated with the problems of the world: crime, corruption, poverty, tragedy, disaster, etc. It's no wonder the majority of our thoughts focus on our problems and things that we don't have. We have been taught to wait until these disasters happen to try to "fix" them or to focus on all the obstacles that lie in the way of achieving our goals.

Your unlimited power lies in your ability to control your thoughts. Remember that most of the circumstances going on in your life and the things around you are an exact mirror of the thoughts that you are allowing to dominate your mind. Simply stated, "*It's all in your head!*"

Being happy comes from the *inside*. It comes from who you are and how you feel about yourself. It has to do with your identity in your own mind. How do you feel about yourself? Do you like who you are? Do you trust yourself? Do you keep promises you make to yourself? Do you have integrity? Do you think you are a good person?

If you are really serious about making big changes in your life, you must start to believe, or better yet, to know that you are great just the way you are. You are a unique person with very special talents.

What We Worry About

Do you realize that we worry about problems that 92 percent of the time won't even manifest? That means we use an immense amount of energy on things that will never occur instead of focusing that energy on solving exciting challenges.

We worry about things in the future all the time. I read a study once that said:

- 40 percent of the time we worry about things that will never happen

- 30 percent of the time we worry about things in the past that we can't change

- 12 percent of the time we worry about things that are illogical—like our chances of getting a certain type of cancer

- 10 percent of the time we worry about small, miscellaneous things

- and only 8 percent do we worry about legitimate worries

That last 8 percent can be broken down even further: 2 percent of our worries we cannot do anything about and 6 percent of our worries are things that we can possibly influence. That means that most of our anticipated problems generally don't occur. Also, the past is the past, so let it go. There isn't anything that you can change about something that already happened even one second ago! Live in the now and have a sense of present time consciousness.

Build Your Resilience

We have all known people who seem to be able to adapt, cope, and respond to stressful circumstances more easily than others. Some people are cool under pressure and able to rise to challenges like hitting their two foul shots to win the game when no time is left on the clock. These people have certain qualities that make them naturally resilient, even when faced with high levels of stress. If you want to build your resilience, start by developing some of these attitudes and behaviors.

- See problems and stressful situations as challenges that can be met.

- View roadblocks and setbacks as temporary and solvable.

- Build strong relationships with family and friends and keep your commitments to them.

- Have a supporting cast and don't be afraid to ask for help.

- Have regular activities that include relaxation and enjoyment.

- Believe that you will succeed, and envision a positive outcome. After all, you can do anything that you set your mind to

- Take substantial steps to solve problems as they arise so they won't get out of hand.

Challenges can be managed as opportunities, and stressors can be seen as challenges, not disasters.

Practice solving problems and asking for help rather than letting fear and anxiety build up. Goal setting and keeping track of your progress are important tools to help you on your way to success. It is important to schedule some time for yourself to unwind and relax. Believe in yourself and be optimistic about the possibilities available to you. Let a little bit of stress motivate you to take positive action toward achieving your goals and realizing your dreams.

Chapter 6

Eliminating Toxins

According to the Environmental Protection Agency (EPA), over 4 billion pounds of toxic chemicals are released by industry into the nation's environment each year, including 72 million pounds of recognized carcinogens (cancer-causing agents). Currently, there are only 650 substances whose release into the environment is tracked by the EPA, which represents less than 1 percent of the more than 75,000 chemicals manufactured in the United States. To make matters worse, the EPA has never systematically reviewed the available environmental health data to determine just how toxic these environmental contaminants are.

Even if we just consider the environmental contamination of the 650 known toxic substances, the numbers are staggering. There are more than 1,300 sites around the country that are so contaminated the EPA has placed them on the Superfund's National Priorities List as critical cleanup sites. Of the 30 hazardous substances found most often at these sites, 18 are known or suspected human carcinogens, and virtually all are associated with noncancer health effects such as toxicities to the liver, kidney, and reproductive systems, including birth defects and low birth weight. There are currently about 11 million people. including 3 to 4 million children, who live within a mile of one of these sites.

Even if we live in a part of the country where we might avoid the industrial contaminants that are present in the big cities, it is almost impossible to do so. Why? Because toxic chemicals are everywhere. They can be in dental fillings, in fluoridated and chlorinated water, in food and food additives, and in personal care products. They can be in the air we breathe and the water we drink. In today's world, we live in a sea of toxins.

Laws are being passed in an effort to protect the environment from these toxins. However, environmental protection laws usually are not enacted until toxic levels of chemicals have already created a great deal of harm to the environment. In addition, new chemicals are being introduced into the marketplace so quickly that adequate safety testing and regulation are rare.

Science continues to discover new health threats from existing chemicals, such as endocrine system problems from estrogen-mimicking pesticides.

Since 1976 the EPA has been conducting the National Human Adipose Tissue Survey (NHATS), a program that collects and chemically analyzes samples of adipose (fat) tissue for the presence of toxic compounds. The objective of the program is to detect and quantify the prevalence of toxic compounds in the general population. Specimens are collected from cadavers and elective surgeries all over of the country.

In 1982 the EPA expanded beyond their normal list to look for the presence of 54 additional chemical toxins. The. results were astounding. Five of these chemicals—OCDD (a dioxin) and four solvents, styrene, 1,4-dichlorobenzene, xylene, and ethylphenol—were found in 100 percent of the human fat tested, including both cadavers and surgical patients. *Everyone tested had all five chemical toxins in their adipose tissue*, often in alarming quantities.

Another nine chemicals were found in 91 to 98 percent of all samples, including such toxins as benzene, toluene, chlorobenzene, ethylbenzene, one furan, three dioxins, and DDE. DDE is formed by a partial dechlorination of DDT, and it can occur in the human body within six months of exposure to DDT. In addition, PCBs were found in 83 percent of the samples, yielding a total of 20 toxic compounds found in 76 percent or more of the samples.

A study of four-year-old children in Michigan revealed the presence of DDT in 70 percent of the subjects, PCB in 50 percent, and PBB in 21 percent.

These ongoing assessments have shown quite clearly that it is not a question of whether we are carrying a burden of toxic compounds. It is a question of which ones, how they interact, and how they affect our health.

How Can We Limit the Toxins in our Bodies?

In your day-to-day life you are exposed to toxins on numerous levels, some within your control, such in food, some totally outside your control, and others in between. If toxicity or deficiency is the basis of disease, then we must understand how to minimize the toxicities. Therefore, it makes sense to learn something about the kinds of exposures that you encounter in order to make informed choices to reduce your overall level of toxic intake.

The 13 Most Common Toxins to Avoid

The following toxins are commonly found in our air, water and food.

1. **PCBs (polychlorinated biphenyls).** This industrial chemical has been banned in the United States for decades, yet it is a persistent organic pollutant that is still present in our environment.

    Risks: Cancer, impaired fetal brain development

    Major Source: Farm-raised salmon. Most farm-raised salmon, which accounts for most of the supply in the United States, are fed meals of ground-up fish that have absorbed PCBs in the environment, and for this reason such salmon should be avoided.

2. **Pesticides.** According to the EPA, 60 percent of herbicides, 90 percent of fungicides, and 30 percent of insecticides are known to be carcinogenic. Alarmingly, pesticide residues have been detected in 50 percent to 95 percent of U.S. foods.

    Risks: Cancer, Parkinson's disease, miscarriage, nerve damage, birth defects, blocking the absorption of nutrients

    Major Sources: Food (fruits, vegetables and commercially raised meats), bug sprays

3. **Mold and other Fungal Toxins.** One in three people have had an allergic reaction to mold. Mycotoxins (fungal toxins) can cause a range of health problems with exposure to only a small amount.

    Risks: Cancer, heart disease, asthma, multiple sclerosis, diabetes

    Major Sources: Contaminated buildings, food like peanuts, wheat, corn and alcoholic beverages

4. **Phthalates.** These chemicals are used to lengthen the life of fragrances and soften plastics.

    Risks: Endocrine system damage (phthalates chemically mimic hormones and are particularly dangerous to children)

    Major Sources: Plastic wrap, plastic bottles, plastic food storage containers. All these can leach phthalates into our food.

5. **Volatile Organic Compounds (VOCs).** These are a major contributing factor to ozone, an air pollutant. According to the EPA, VOCs tend to be even higher (two to five times) in indoor air than outdoor air, likely because they are present in so many household products.

    Risks: Cancer, eye and respiratory tract irritation, headaches, dizziness, visual disorders, and memory impairment

    Major Sources: Drinking water, carpet, paints, deodorants, cleaning fluids, varnishes, cosmetics, dry cleaned clothing, moth repellants, air fresheners.

6. **Dioxins.** Chemical compounds formed as a result of combustion processes such as commercial or municipal waste incineration and from burning fuels (like wood, coal or oil).

    Risks: Cancer, reproductive and developmental disorders, chloracne (a severe skin disease with acne-like lesions), skin rashes, skin discoloration, excessive body hair, mild liver damage.

    Major Sources: Animal fats are responsible for more than 95 percent of exposure.

7. **Asbestos.** This insulating material was widely used from the 1950s to 1970s. Problems arise when the material becomes old and crumbly, releasing fibers into the air.

    Risks: Cancer, scarring of the lung tissue, mesothelioma (a rare form of cancer)

    Major Sources: Insulation on floors, ceilings, water pipes and healing ducts from the 1950s to the 1970s.

8. **Heavy Metals.** Metals like arsenic, mercury, lead, aluminum and cadmium, which are prevalent in many areas of our environment,

can accumulate in soft tissues of the body. Nearly all organ systems are involved in heavy metal toxicity, but the most common systems are the nervous, gastrointestinal, cardiovascular, and renal systems. The EPA produced a study in 1999 stating that toxic metals are the second worst environmental health problem in the United States.

Aluminum. Antacids, OTC drugs, douches, cookware, foil, antiperspirants, baking powder, water, food additives, margarine processing.

Arsenic. Soil, seafood, fuel, oils, coal, weed killer, pesticides, water, laundry aids, tobacco.

Cadmium. Tobacco, refined foods, water pipes, coal burning, coffee and tea, shellfish, soft drinks, fungicides, pesticides, plastics.

Copper. Water, cookware, oral contraceptives, copper supplements, natural food (whole grains, shellfish, liver, beans, nuts).

Lead. Gasoline, paint, foods, water, pottery, cans, cosmetics, cigarettes, pesticides, liver, air pollution.

Mercury. Pesticides, cosmetics, dental fillings, seafood, medicines, laxatives, inks, tattoos, paint.

Nickel. Food, dental metals, jewelry, air pollution, tobacco smoke, car exhaust, industrial waste, cooking utensils, hydrogenated fats, fertilizers.

Risks: Cancer, neurological disorders, Alzheimer's disease, Attention Deficit Disorder (ADD), foggy head, fatigue, nausea and vomiting, decreased production of red and white blood cells, abnormal heart rhythm, damage to blood vessels, osteoporosis.

Major Sources: Drinking water, fish, vaccines, pesticides, preserved wood, antiperspirant, building materials, dental amalgams, chlorine plants.

9. **Chloroform.** This colorless liquid has a pleasant, nonirritating odor and a slightly sweet taste, and is used to make other chemicals. It is also formed when chlorine is added to water.

Risks: Cancer, potential reproductive damage, birth defects, dizziness, fatigue, headache, liver and kidney damage.

Major Sources: Air, drinking water and food can contain chloroform.

10. **Chlorine.** This highly toxic yellow-green gas is one of the most heavily used chemical agents.

    Risks: Sore throat, coughing, eye and skin irritation, rapid breathing, narrowing of the bronchi, wheezing, blue coloring of the skin, accumulation of fluid in the lungs, pain in the lung region, severe eye and skin burns, lung collapse, reactive airways dysfunction syndrome (RADS, a type of asthma).

    Major Sources: Household cleaners, drinking water (in small amounts), air when living near an industry such as a paper plant that uses chlorine in industrial processes.

11. **Fluoride.** This highly toxic gas is added in municipal water systems in an effort to decrease dental problems in children. An 11-year study of 39,000 schoolchildren showed no statistically significant difference in tooth decay between those using fluoride and those who didn't. The study also found that fluoride damaged brain enzymes and lowered IQ. *Fluoride 1996, Zhao*

    Risks: Disrupted brain and liver enzymes, and lowered IQ.

    Major Sources: Drinking water and Teflon coatings on cookware.

12. **Electromagnetic Toxicity.** Although this is not a physical compound, there is considerable evidence that our exposure to the electromagnetic fields produced by everything from power lines to cell phones can have a detrimental affect on health. A two-year study by the FDA on extremely low-frequency fields (ELFs) recommended that these fields be listed as probable human carcinogens, alongside chemicals like PCB's, formaldehyde, and dioxin.

    Risks: Increase cancer risk, disruption of normal mental functioning.

    Major Sources: Household appliances, electrical devices, computers, cell phones radios and other electrical devices.

13. **Tobacco Products and Secondhand Smoke.** There are 4000 chemicals in tobacco with 100 identified poisons and 63 known drugs.

According to the American Cancer Society, secondhand smoke can cause 35,000 deaths from heart disease each year, 3000 lung cancer deaths each year, increased number and severity of asthma attacks in children, and increased number of cases of children with fluid and inflammation of the middle ear. Infants who are exposed to secondhand smoke have an increased risk of sudden infant death syndrome (SIDS).

Risks: Lung diseases, cancer, impaired circulation, heart disease, impaired immunity, degenerative disc disease, middle ear infections.

Major Sources: Work places, public places, homes.

How Do These Toxins Affect Your Health?

The twentieth century, with its promise of better living through chemistry, instead generated a host of chemical toxins and related illnesses, often referred to as environmental illnesses. Recent articles in the medical literature have shown that the rate of cancers not associated with smoking are higher for those born after 1940 than before, and that this increase in the cancer rate is due to environmental factors other than smoking.

New medical diagnoses include sick (closed) building syndrome and multiple chemical sensitivities (MCS), both of which are known to be related to overexposure to environmental contaminants.

The primary action of the major pesticide classes and solvents is to disrupt neurological function. In addition to being neurotoxic, these compounds are extremely toxic to the immune and endocrine systems. The adverse health effects are not limited only to those systems; these compounds can cause a variety of dermatological, gastrointestinal, genitourinary, respiratory, musculoskeletal, and cardiac problems as well.

Toxin-Associated Cancers

Several hundred of the 75,000 chemicals currently being released into the environment are known or suspected carcinogens. These chemicals interfere with the normal cell growth and development, resulting in areas of tissue that grow out of control. The more toxic exposure an individual has, the greater the risk of developing a cancer.

There is a high correlation between toxic chemical exposure and the rate of cancer in the population. Men born in the 1940s had twice the cancer incidence as those born from 1888-1897, even when smoking was factored out. Women born in the 1940s had 50 percent more total cancers, and 30 percent more white women in particular had cancer not linked to smoking. Cancers in children are at an all-time high, especially brain cancers in children who have been exposed to pesticides by means of flea and tick collars on the family pet, termite treatments, flea bombs, and other sources. Several studies have shown that leukemias and myelomas are associated with exposures to environmental toxins, especially industrial solvents.

Because many of the carcinogenic toxins resemble estrogen and other hormones in the body, cancers tend to form in tissues that are sensitive to these hormones, hence the dramatic increase in breast cancer, testicular cancer, and ovarian cancer.

Neurotoxicity

Neurotoxicity simply means that a particular compound is toxic to the nervous system. Most of the major classes of pesticides are neurotoxins by design—they kill pests by attacking the nervous system. Numerous studies have shown that exposure to pesticides and agricultural fertilizers contribute to the development of Alzheimer's disease, Parkinson's disease, and multiple sclerosis (MS), as well as neuropathies in the hands and feet.

Even more disturbing is that some of the compounds that have been shown to be neurotoxic in the laboratory are actually used as food additives, such as the flavor-enhancers monosodium glutamate (MSG) and aspartame (NutraSweet). These compounds are known as *excitoneurotoxins*, meaning that their toxic effects are caused by overstimulating the nerves. Attention deficit disorder, convulsions, obesity, and learning disorders have all been linked to consumption of chemical flavoring agents in the laboratory.

Detoxing the Toxins

The body has two major detoxification systems: the antioxidation system and the cytochrome P-450 enzyme complex in the liver. They both work in conjunction with the body's circulatory and elimination systems. Even a superficial understanding of how these systems work can give you a whole new basis of health, because with this understanding, you can improve your body's ability to detoxify, based on scientific knowledge. You will understand how and why these measures help to protect your health.

Many toxins that we consume or are exposed to will not dissolve in water. Called lipid-soluble, they only dissolve in oil or oily solutions. Fatty tissue, or tissues that have lipid soluble membranes, like the liver, can store lipid soluble toxins for months or years. The job of the liver is to convert these toxins from lipid-soluble to water-soluble. Once they can be dissolved by water, they can be flushed from the body by the kidneys or the bowel. This change takes place in the liver through a complex system of enzymes.

The liver has the ability to break down and remove other invaders as well, such as debris, bacteria, and chemical toxins, but a liver that is overloaded with toxins may not be able to keep up with the demand for its services. When this is the case, it will continue to store toxins, even if storage of those toxins over a long period of time might result in liver damage. Supporting the liver involves two steps: limiting the amount of toxins you take in and aiding the liver in processing previous toxic exposure.

The liver is a primary player in your body's immune system. Supporting your liver, limiting your exposure to toxins, undertaking liver flushes, and increasing intake of healthy fluids can be important steps in supporting healthy immune function and responding to toxin-related illnesses.

A Basic Detoxifying Program

The purpose of detoxifying (or detox) is to neutralize and eliminate any compound in the body that can be toxic. Detox is a natural process occurring on a continual basis in the body, but because of the modern diet, the enormous number of chemicals we ingest daily, and the increase in chronic degenerative diseases, many people find that a regimen to help the body detox more effectively is necessary. Such a program strengthens the organs involved in detoxification, such as your liver, and helps to release and eliminate stored toxins.

Liver Detoxification

The liver is responsible for breaking down or transforming substances like ammonia, metabolic waste, drugs, alcohol, and chemicals, so that they can be excreted. Inside the liver cells there are sophisticated mechanisms that to break down toxic substances. Every drug, artificial chemical, pesticide, and hormone is broken down (metabolized) by enzyme pathways inside the liver cells. Many of the toxic chemicals that enter the body are fat-soluble, so they dissolve only in fatty or oily solutions and not in water. This makes them difficult for the body to excrete.

Most liver detox formulas on the market are based around an herb called milk thistle, which contains a compound called Silymarin. This compound stimulates the detoxifying enzyme systems in the liver, of which the most important is the cytochrome P-450 enzyme complex. I take Milk Thistle every two months, and I recommend it to my patients. Several amino acids such as glutamine, glutathione, taurine and glycine, are also important to help the liver effectively break down toxic substances.

Heavy Metal Detoxification

Scientists estimate that more than half of all Americans have too much heavy metal in their bodies. These metals, which include lead, mercury, aluminum, cadmium, and arsenic, are found in industrial byproducts and in tainted drinking water, pesticides, and even dental fillings and cooking utensils. Heavy metal poisoning is very difficult to treat and mostly unrecognized, but widely common. Many conditions related to the nervous system are directly linked to heavy metal poisoning.

It has been known for some time that the removal of these heavy metals from the body can result in a substantial improvement in health. Until very recently, the process of removing these heavy metals, called chelation, had to be done in a medical setting, due to the harsh nature of the chelating agents used.

It was discovered a few years ago that the common herb cilantro has strong chelating properties and is much less stressful on the body. Since then, a number of studies have shown that compounds in cilantro help to bind heavy metals and remove them from the body and even from the brain.

You should consider going through a heavy metal detox every few years to keep the inside of your body clean and healthy. There are a number of heavy metal detox products on the market to choose from. For more information, consult your wellness chiropractor.

Candida Detoxification

Candidiasis, an infection caused by candida fungi, is commonly called a yeast infection. These fungi are found almost everywhere in the environment. Some may live harmlessly along with the abundant native species of bacteria normally found in the body. Usually, candida is kept under control by the native bacteria and by the body's immune defenses. If the native bacteria are decreased by antibiotics or weakened by illness (especially AIDS or diabetes), malnutrition, consuming too much processed sugar, or taking certain medications, candida fungi can multiply and cause significant problems.

The key to combating candida is to decrease the amount of processed sugar you consume and to take supplemental probiotics. An expert panel commissioned by the World Health Organization defined probiotics as "Live microorganisms which, when administered in adequate amounts, confer a health benefit on the host." Our ancestors had a diet brimming with bacteria. Food was usually eaten raw or fermented, introducing billions of bacteria into the gut with every meal. Plants were particularly abundant, and stone age people consumed over ten times as many varieties as we do these days.

Several decades of oversterilization of the environment have denied many of us our needed exposure to microorganisms in the soil, resulting in insufficient immune systems that have not been properly educated. The intestinal tract should normally have a bacteria balance of 85 percent beneficial to 15 percent harmful, but today, most of us have the reverse ratio!

Several probiotic supplements are available on the market. For information on which one may be best for you, talk to your wellness chiropractor.

Colon Cleanse

Colon cleansing uses natural healing and herbs to help your body to heal health problems related to a colon that is not functioning as nature intended. Many people have problems with the colon and don't realize it. Poor diet and exposure to environmental toxins causes a buildup of toxic waste in the colon. In addition to the poisonous effect this has on your body, waste buildup hinders nutrient absorption. By not eating healthy foods and not exercising on a regular basis, your system has a tough time digesting foods properly.

Colon cleansing programs are based on increased fiber intake that can reduce symptoms of diverticulosis and prevent complications such as diverticulitis. Fiber keeps stool soft and lowers pressure inside the colon so that bowel contents can move through easily.

For colon cleansing it is best to use a fiber supplement designed specifically for this purpose. Good quality supplements contain both soluble and insoluble fiber and are formulated to loosen the stool, absorb and sweep away fat and toxins, reduce transit time, and make elimination effortless and complete. Colonics may also be appropriate.

Tips for Avoiding Toxins

It's impossible to avoid all environmental toxins, but you can limit your exposure by following a few tips.

- As much as possible, eat organic produce and free-range, organic foods.

- Rather than eating fish, which is largely contaminated with PCBs and mercury, consume a high-quality purified fish oil. Go to www.innatechoice.com for more information.

- Avoid processed foods—remember, they're processed with chemicals!

- Use only natural cleaning products in your home

- Switch to natural brands of toiletries

- Remove any metal fillings, as they're a major source of mercury. Be sure to have this done by a qualified dentist.

- Avoid artificial air fresheners, dryer sheets, fabric softeners or other synthetic fragrances that are often harmful to your skin and to the air around you.

- Avoid artificial food additives, including artificial sweeteners and MSG

- Get plenty of safe sun exposure to boost your vitamin D levels and your immune system

- Have your tap water tested. If contaminants are found, install an appropriate water filter on all your faucets.

- Avoid using aluminum cookware and aluminum foil for cooking and storing food. Choose glass or high-grade stoneware instead.

- Avoid tobacco products and secondhand smoke.

Ten Common Toxicities

Processed foods
Prescription medications
Smoking
Microwaved food
Aspartame, MSG and Splenda
Homogenized and pasteurized dairy products
High fructose corn syrup
White sugar
Sun block
Stress from a toxic environment, mental or chemical

How to Prevent 10 Common Deficiencies

Eat more fruits and vegetables, preferably organic
Consume Omega 3 fatty acids in food and/or supplements
Get natural sunlight
Take a whole food supplement daily
Drink enough water
Consume vitamin E in food and/or supplements
Oxygenate your body through exercise
Get plenty of sleep
Keep a positive attitude
Eat complete proteins

Remember, toxicity and deficiencies are the basis of all sickness and disease. I hope this book has enlightened you on how to become healthier and happier than ever before. Follow the principles in this book, and be sure to talk to your wellness chiropractor if you have any questions. Thank you for reading and for taking the first step toward *Health and Wellness ... available NOW without a Prescription.*

About the Author

Dr. Greg S. Tomalin

Dr. Tomalin has owned practices in Broomfield, Westminster, Longmont and Lakewood Colorado. He is also the owner of AmeriSpine Chiropractic. These clinics are some of the fastest growing chiropractic offices in the Denver Metropolitan area. He is the co-author of the national bestselling books *The World's Best Kept Health Secret Revealed* and *Stay Fit While You Sit—Deskersize Your Way to Better Health!* He has published numerous articles in newspapers, magazines and professional journals. He is the co-host of a radio show called Today's Health that can be heard on local Denver radio. He has appeared on the national talk show Phenomenal Health with host Dr. Jeff Hockings. His energy and casual speaking style have made him a popular speaker for a wide variety of seminars and workshops ranging from business groups to social organizations.

Dr. Tomalin received his Bachelor of Arts in psychology from Carleton University in Ottawa, Canada, and a Bachelor of Science in human biology as well as his Doctor of Chiropractic from Logan College in St. Louis, Missouri. He lives with his two beautiful children, Jaden and Kaiya. Dr. Tomalin is a pilot and enjoys riding his Harley-Davidson motorcycle.

Dr. Tomalin offers entertaining presentations and workshops on stress, exercise, nutrition, ergonomics, health, wellness, at the workplace.

If you are interested in having Dr. Tomalin speak at one of your events, please contact him at the following:

Health and Wellness Chiropractic Center/ AmeriSpine Chiropractic
www.DrTomalin.com or drtomalin@gmail.com
720-887-0624 303-919-5477

www.ingramcontent.com/pod-product-compliance
Lightning Source LLC
Chambersburg PA
CBHW020244290526
45784CB00003B/1100